SCAN THE CODE TO ACCESS YOUR FREE DIGITAL COPY OF THE NEUROANATOMY COLORING BOOK

SCAN ME

The Neuroanatomy Coloring Book features:

• **The most effective way to skyrocket your neuroanatomical knowledge, all while having fun!**

• Full coverage of the major systems of the human brain to provide context and reinforce visual recognition

• **25+ unique, easy-to-color pages of different neuroanatomical sections with their terminology**

• Large 8.5 by 11-inch single side paper so you can easily remove your coloring

• **Self-quizzing for each page, with convenient same-page answer keys**

THIS BOOK BELONGS TO

_____

TABLE OF CONTENTS

YOGA POSES FOR EXPERTS

YOGA POSES FOR EXPERTS

1. FULL LORD OF THE FISHES POSE

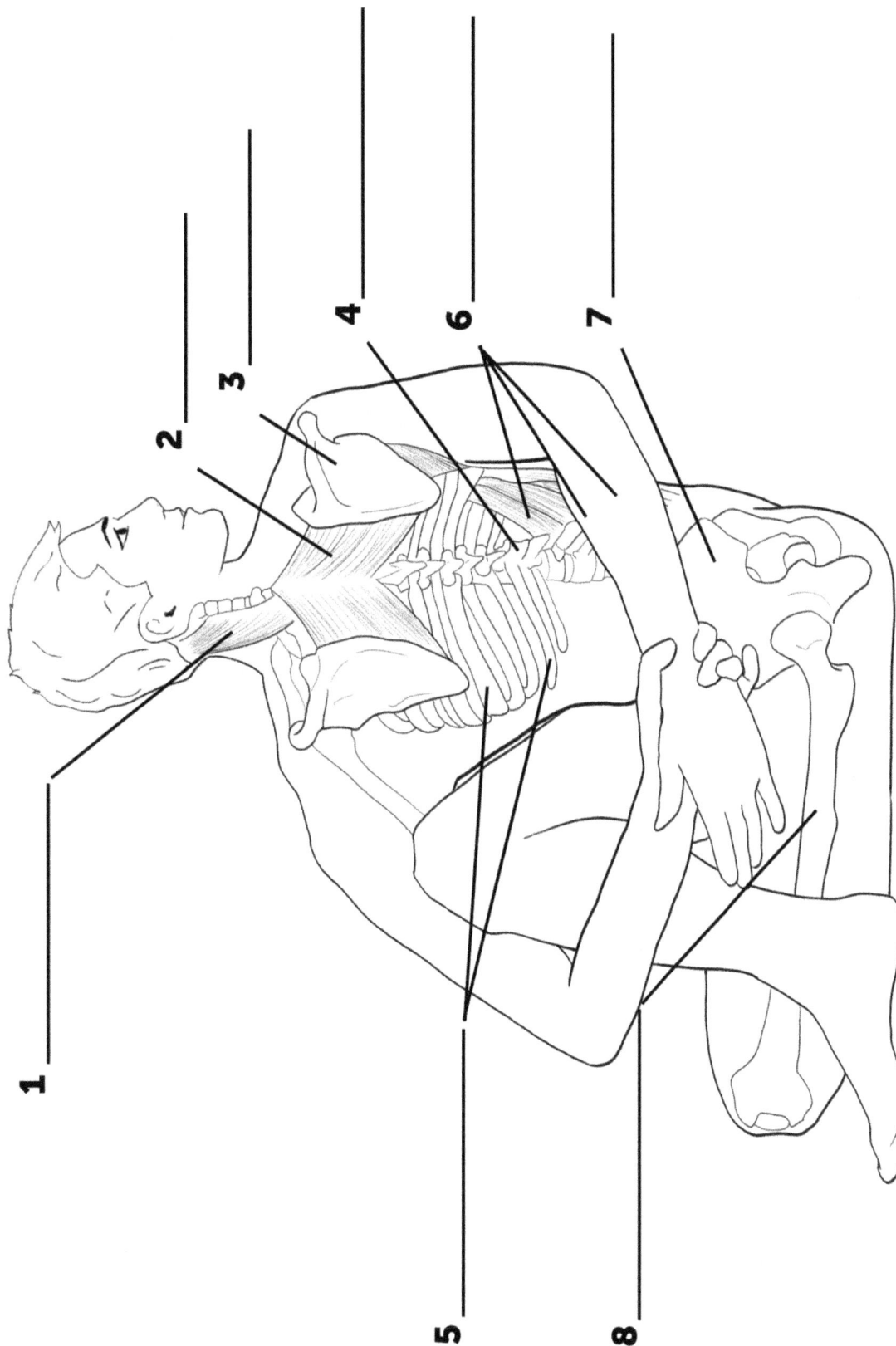

1

2

3

4

5

6

7

8

1. FULL LORD OF THE FISHES POSE

1. SPLENIUS CAPITIS
2. RHOMBOIDS
3. SCAPULA
4. SPINE
5. RIBS
6. ERECTOR SPINAE
7. PELVIS
8. FEMUR

2. FLYING CROW POSE

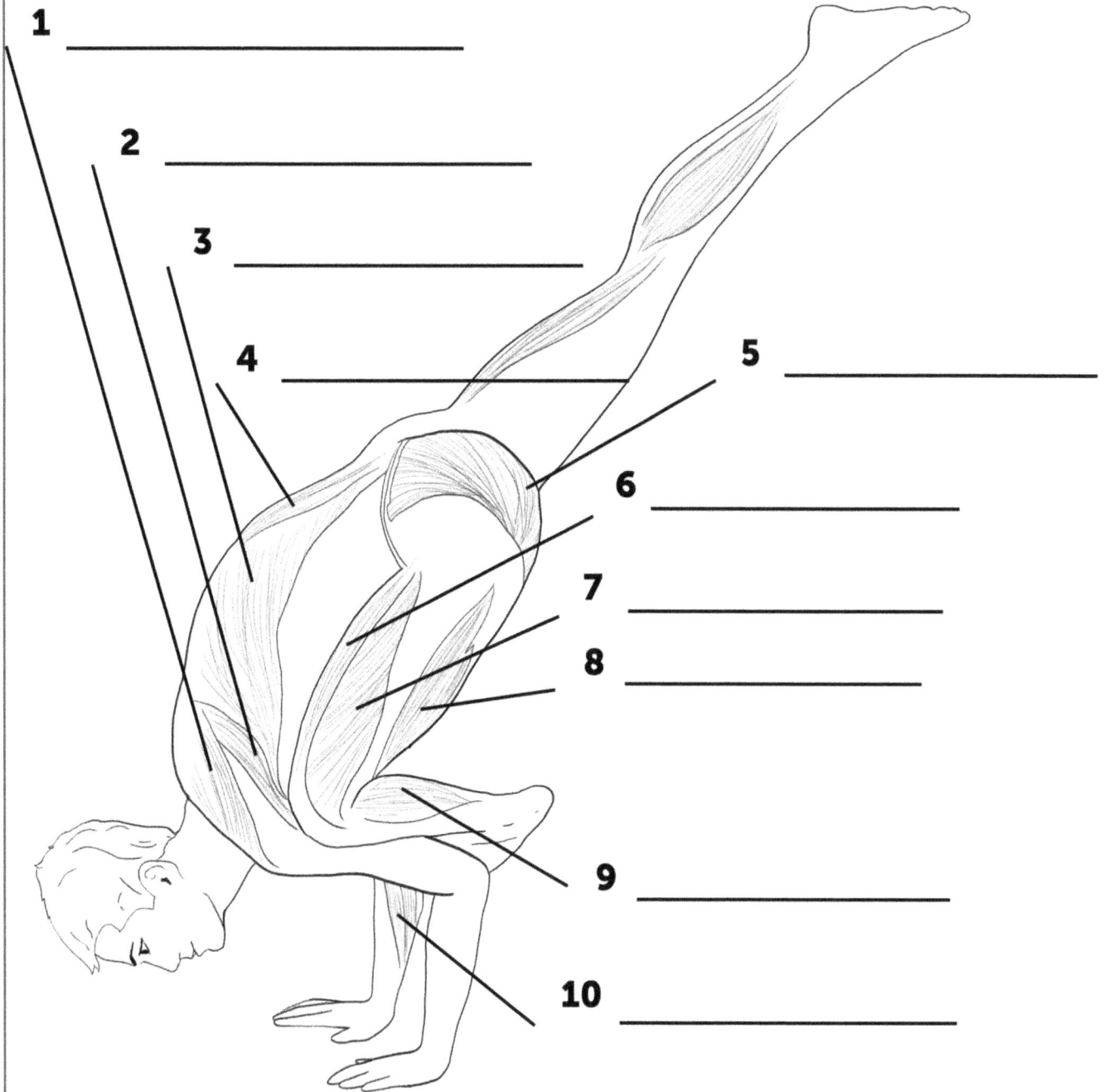

1 _____

2 _____

3 _____

4 _____

5 _____

6 _____

7 _____

8 _____

9 _____

10 _____

2. FLYING CROW POSE

1. DELTOID
2. TRICEPS BRACHII
3. LATISSIMUS DORSI
4. ERECTOR SPINAE
5. GLUTEUS MAXIMUS
6. RECTUS FEMORIS
7. VASTUS LATERALIS
8. HAMSTRINGS
9. GASTROCNEMIUS
10. PRONATORS

3. SCORPION POSE

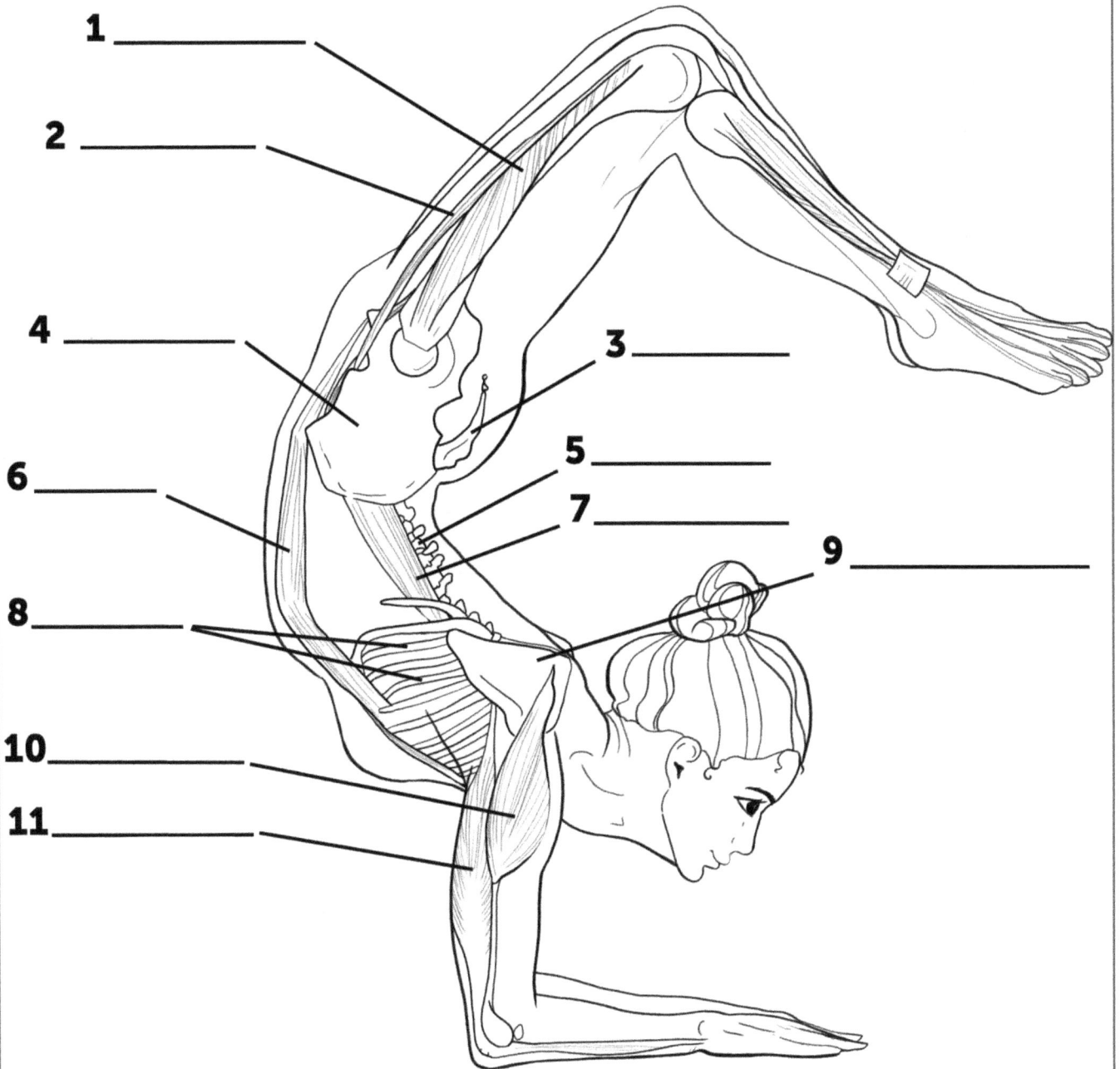

1 _____

2 _____

4 _____

3 _____

5 _____

6 _____

7 _____

9 _____

8 _____

10 _____

11 _____

3. SCORPION POSE

1. VASTUS LATERALIS
2. RECTUS FEMORIS
3. SACRUM BONE
4. PELVIS
5. SPINE
6. RECTUS ABDOMINIS
7. PSOAS MAJOR
8. RIBS
9. SCAPULA
10. DELTOID
11. TRICEPS BRACHII

4. FIREFLY POSE

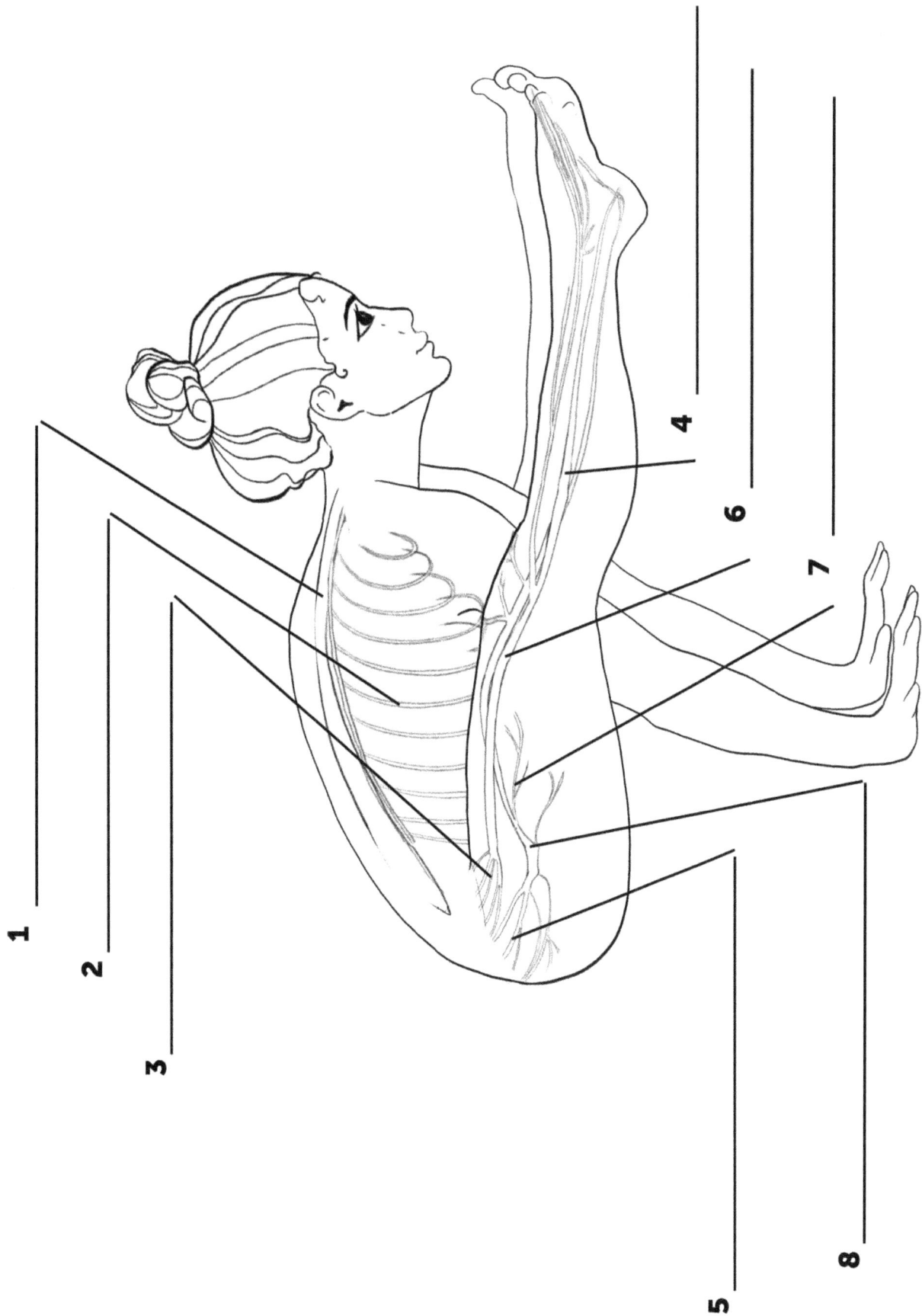

1

2

3

4

5

6

7

8

4. FIREFLY POSE

1. SPINAL CORD
2. INTERCOSTALS
3. SACRAL PLEXUS
4. TIBIAL
5. LUMBAR PLEXUS
6. SCIATIC
7. MUSCULAR BRANCHES OF FEMORAL
8. FEMORAL

5. BIRD OF PARADISE POSE

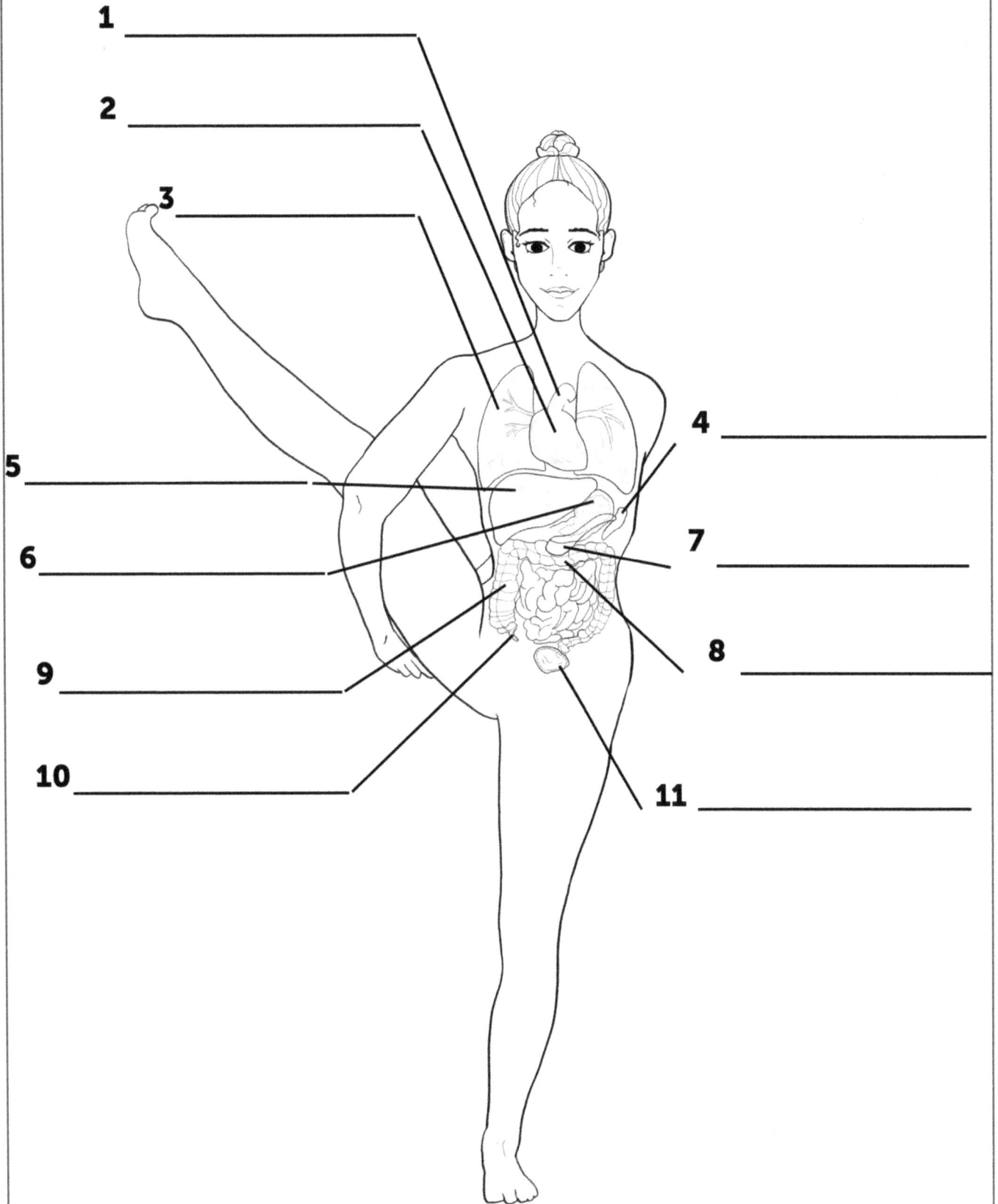

1 _____

2 _____

3 _____

4 _____

5 _____

6 _____

7 _____

8 _____

9 _____

10 _____

11 _____

5. BIRD OF PARADISE POSE

1. AORTA
2. HEART
3. LUNGS
4. SPLEEN
5. LIVER
6. STOMACH
7. PANCREAS
8. TRANSVERSE COLON
9. ASCENDING COLON
10. APPENDIX
11. URINARY BLADDER

6. PEACOCK POSE

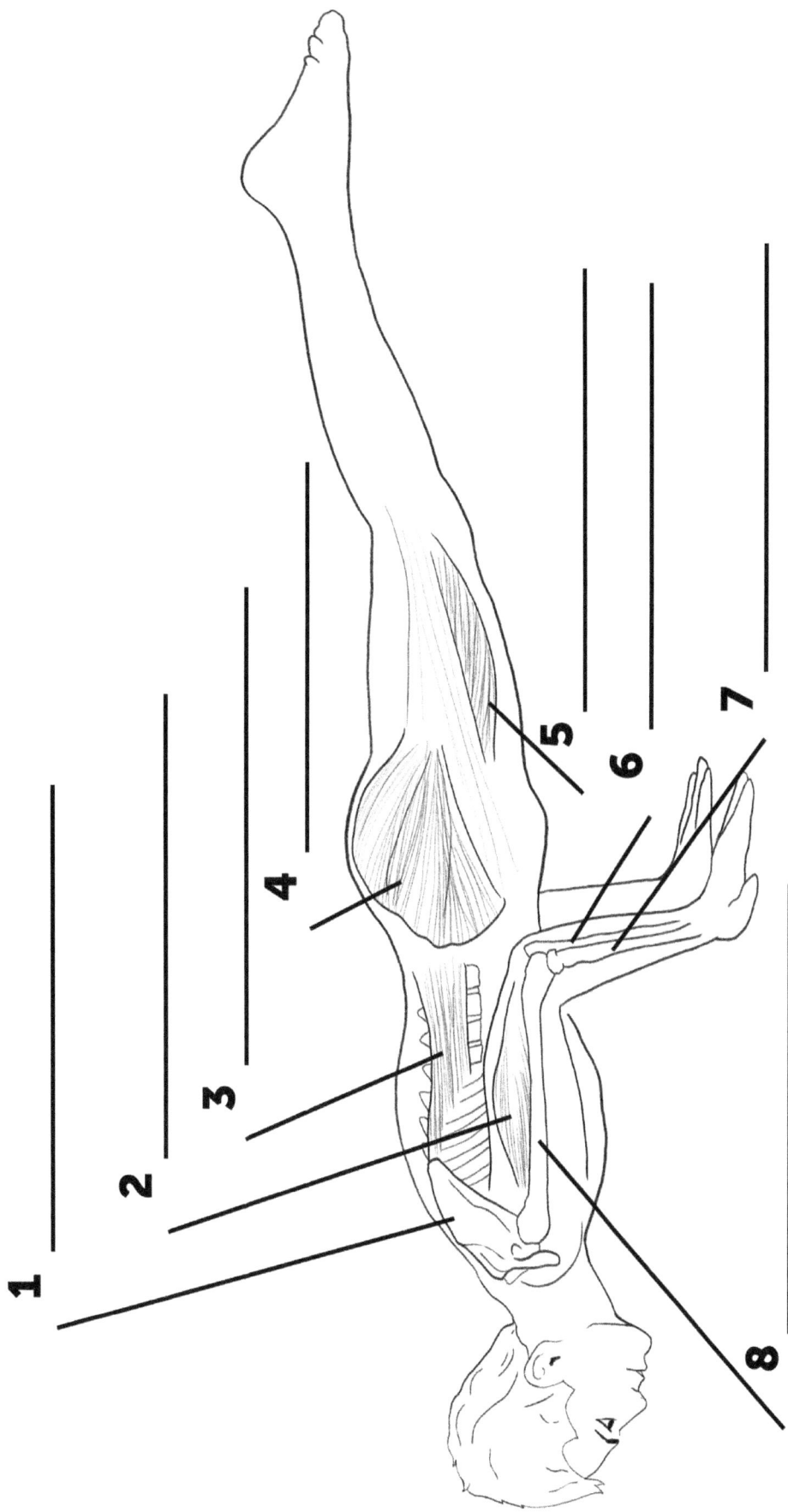

1

2

3

4

5

6

7

8

6. PEACOCK POSE

1. SCAPULA

2. TRICEPS BRACHII

3. ERECTOR SPINAE

4. GLUTEUS MAXIMUS

5. QUADRICEPS

6. ULNA

7. RADIUS

8. HUMERUS

7. ONE-LEGGED KING PIGEON POSE II

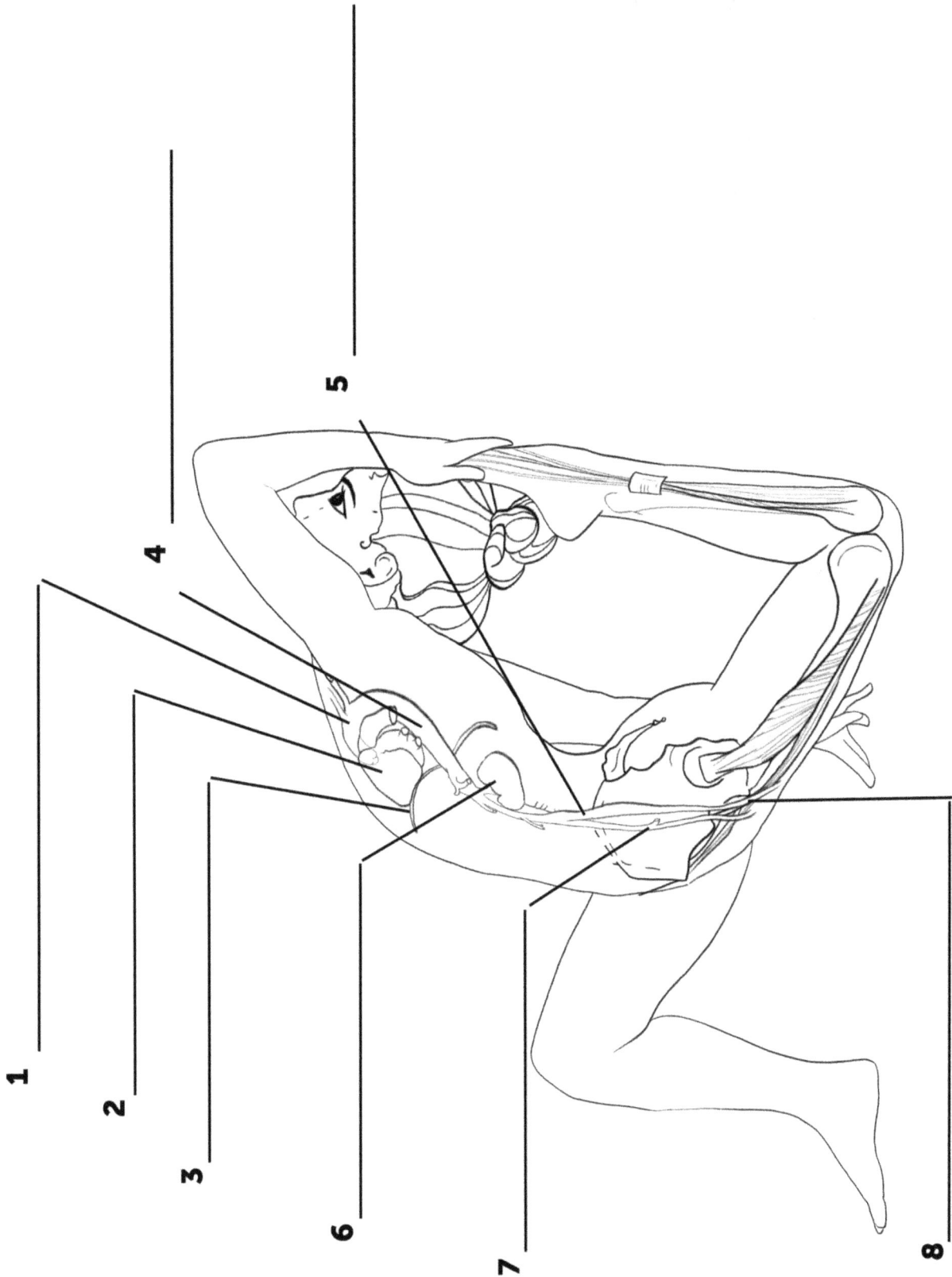

1

2

3

4

5

6

7

8

7. ONE-LEGGED KING PIGEON POSE II

1. ASCENDING THORACIC AORTA

2. HEART

3. DIAPHRAGM

4. DESCENDING THORACIC AORTA

5. ABDOMINAL AORTA

6. KIDNEY

7. COMMON ILIAC ARTERY

8. FEMORAL ARTERY

8. LITTLE THUNDERBOLT POSE

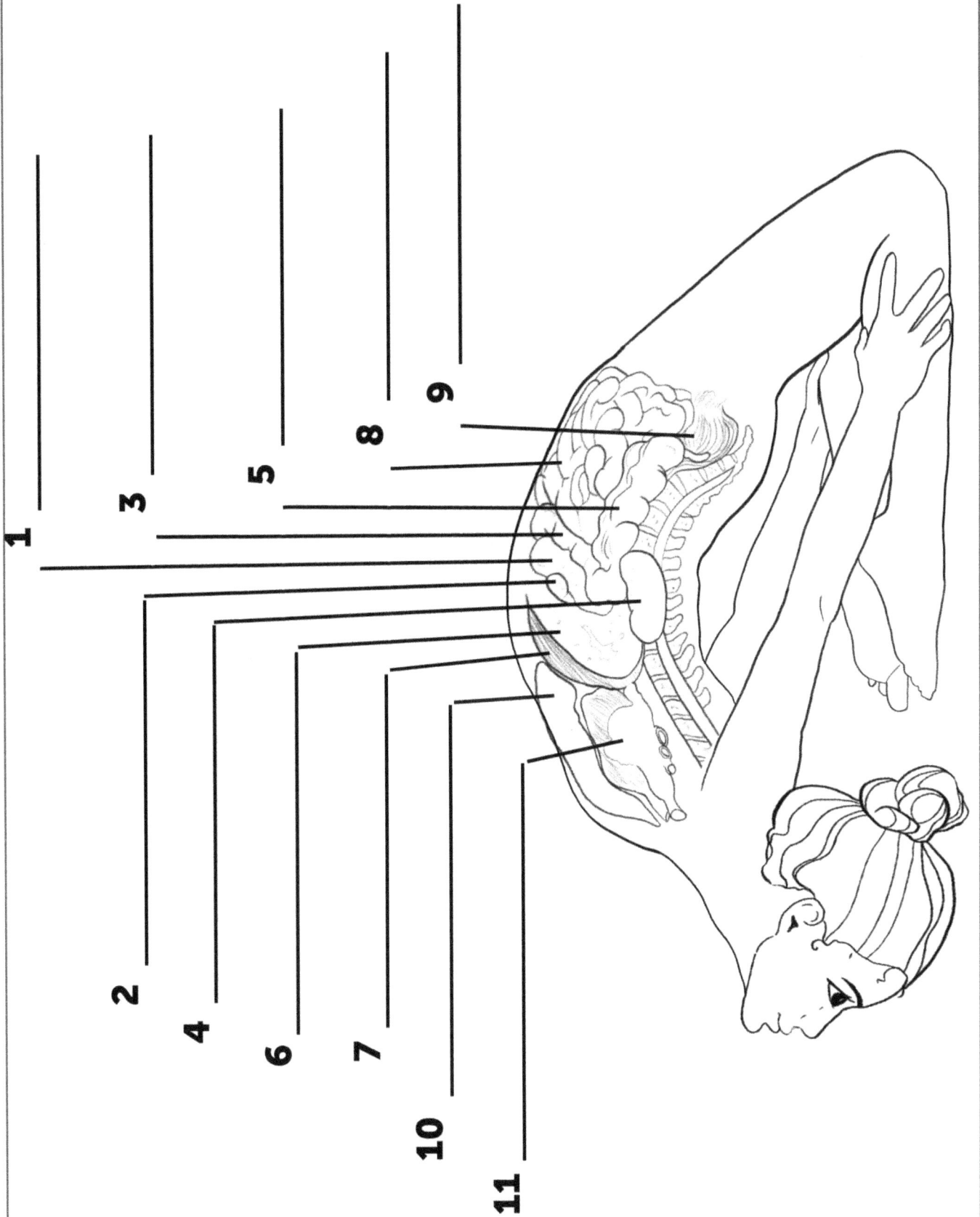

1

2

3

4

5

6

7

8

9

10

11

8. LITTLE THUNDERBOLT POSE

1. STOMACH
2. GALLBLADDER
3. TRANSVERSE COLON
4. KIDNEY
5. ASCENDING COLON
6. LIVER
7. DIAPHRAGM
8. COILS OF SMALL INTESTINE
9. RECTUM
10. LUNGS
11. HEART

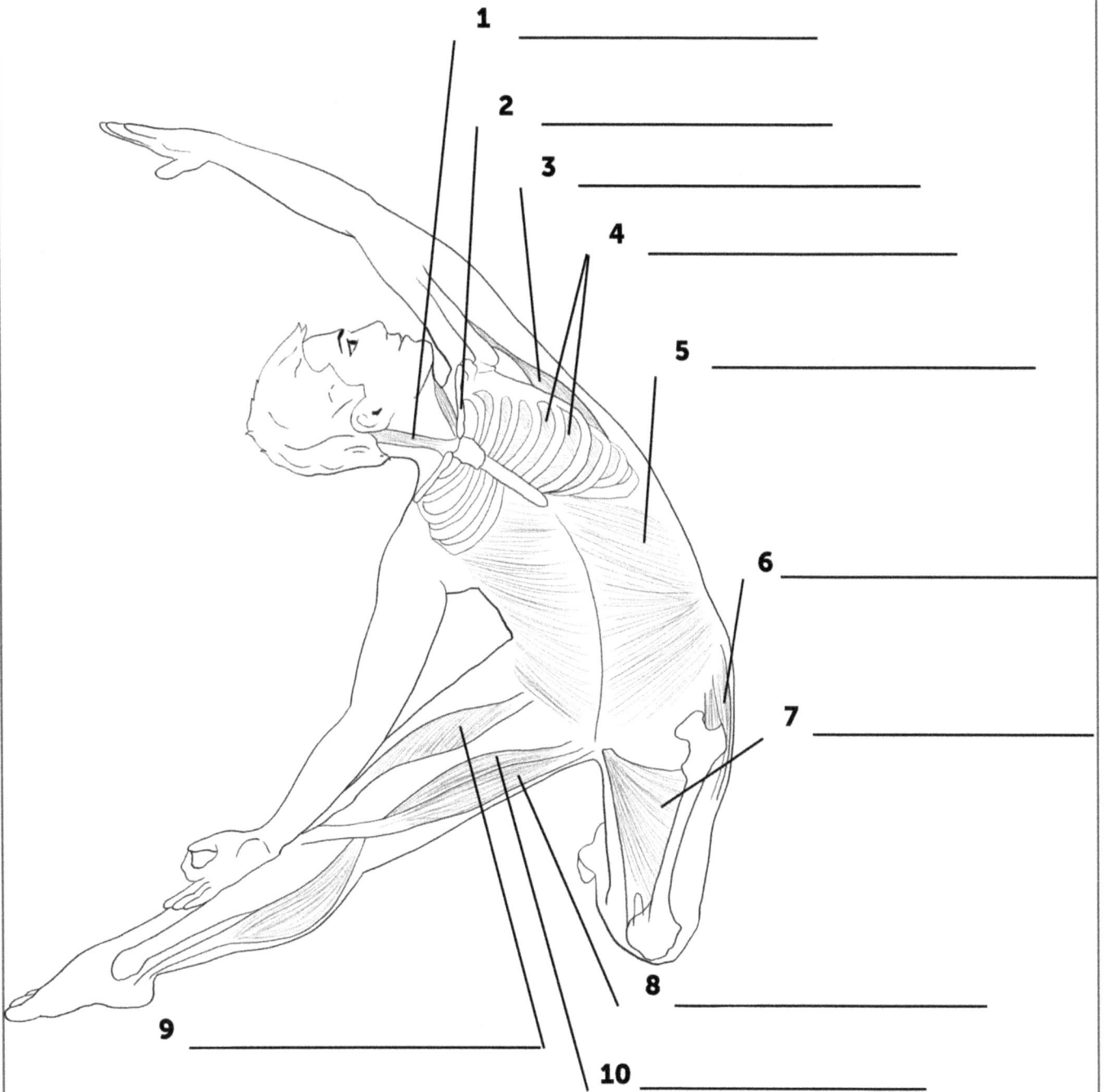

9. GATE POSE

1 _____

2 _____

3 _____

4 _____

5 _____

6 _____

7 _____

8 _____

9 _____

10 _____

9. GATE POSE

1.   SPLENIUS CAPITIS

2.   COLLARBONE

3.   LATISSIMUS DORSI

4.   INTERCOSTALS

5.   EXTERNAL OBLIQUE

6.   TENSOR FASCIAE LATAE

7.   ADDUCTOR LONGUS

8.   GRACILIS

9.   RECTUS FEMORIS

10. ADDUCTOR MAGNUS

10. SAGE KOUNDIYA I POSE

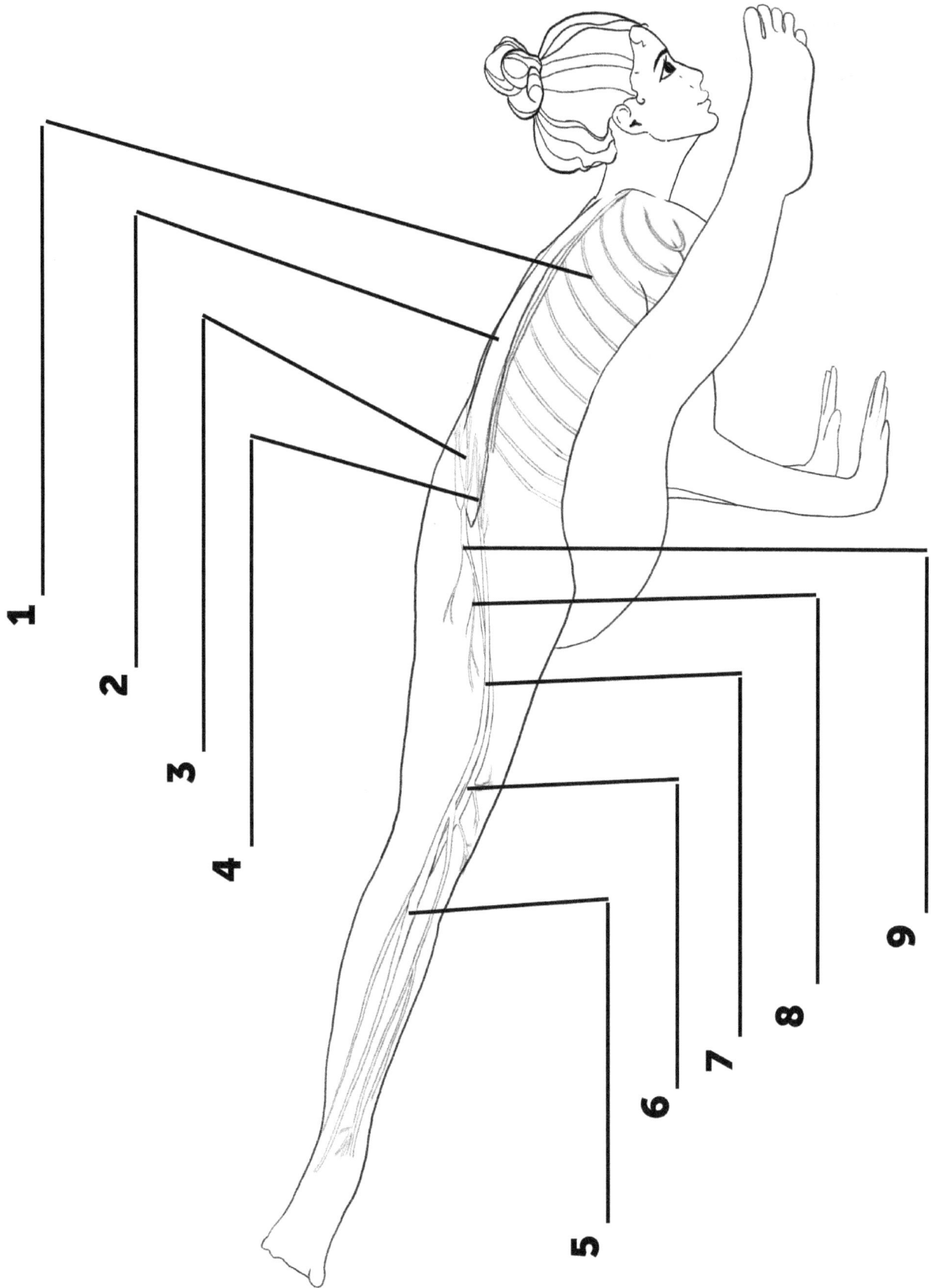

1

2

3

4

5

6

7

8

9

10. SAGE KOUNDIYA I POSE

1. INTERCOSTALS
2. SPINAL CORD
3. LUMBAR PLEXUS
4. SACRAL PLEXUS
5. TIBIAL
6. SAPHENOUS
7. SCIATIC
8. MUSCULAR BRANCHES OF FEMORAL
9. FEMORAL

11. SAGE KOUNDIYA II POSE

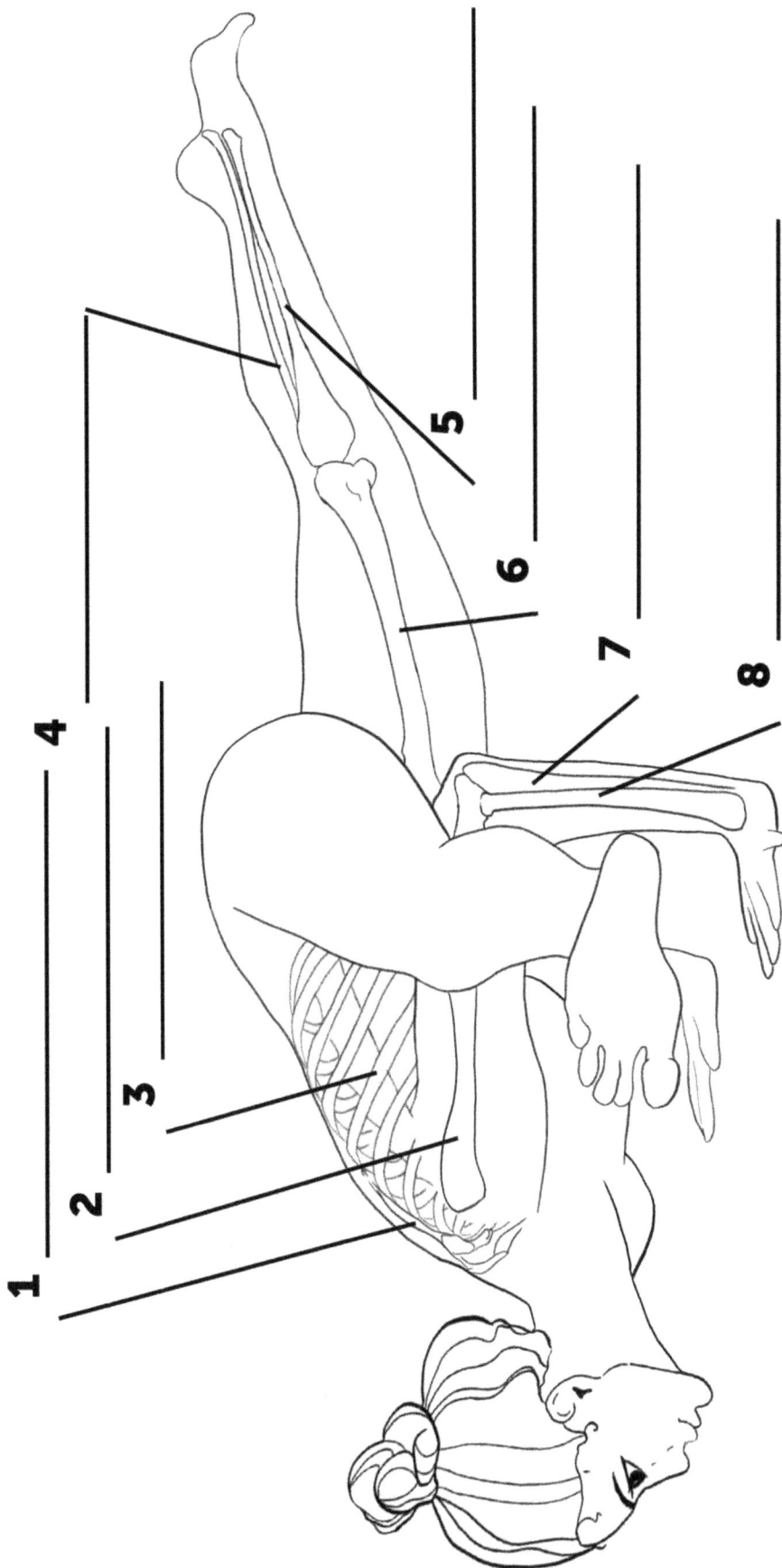

1

2

3

4

5

6

7

8

11. SAGE KOUNDIYA II POSE

1. SCAPULA
2. HUMERUS
3. RIBS
4. FIBULA
5. TIBIA
6. FEMUR
7. ULNA
8. RADIUS

12. HEAD TO FOOT POSE

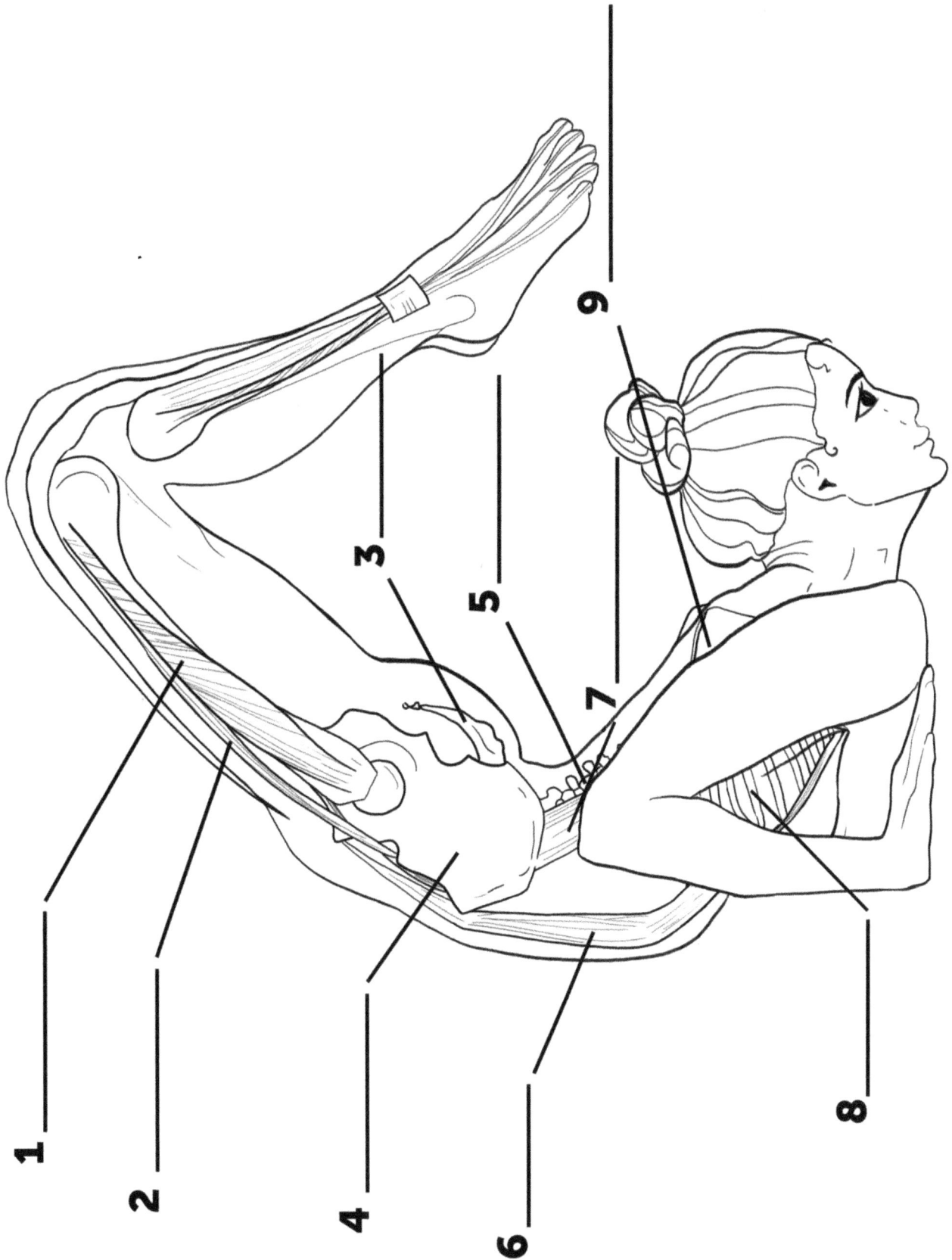

12. HEAD TO FOOT POSE

1. VASTUS LATERALIS

2. RECTUS FEMORIS

3. SACRUM

4. PELVIS

5. SPINE

6. RECTUS ABDOMINIS

7. ERECTOR SPINAE

8. RIBS

9. SCAPULA

13. MASTER BABY GRASSHOPPER POSE

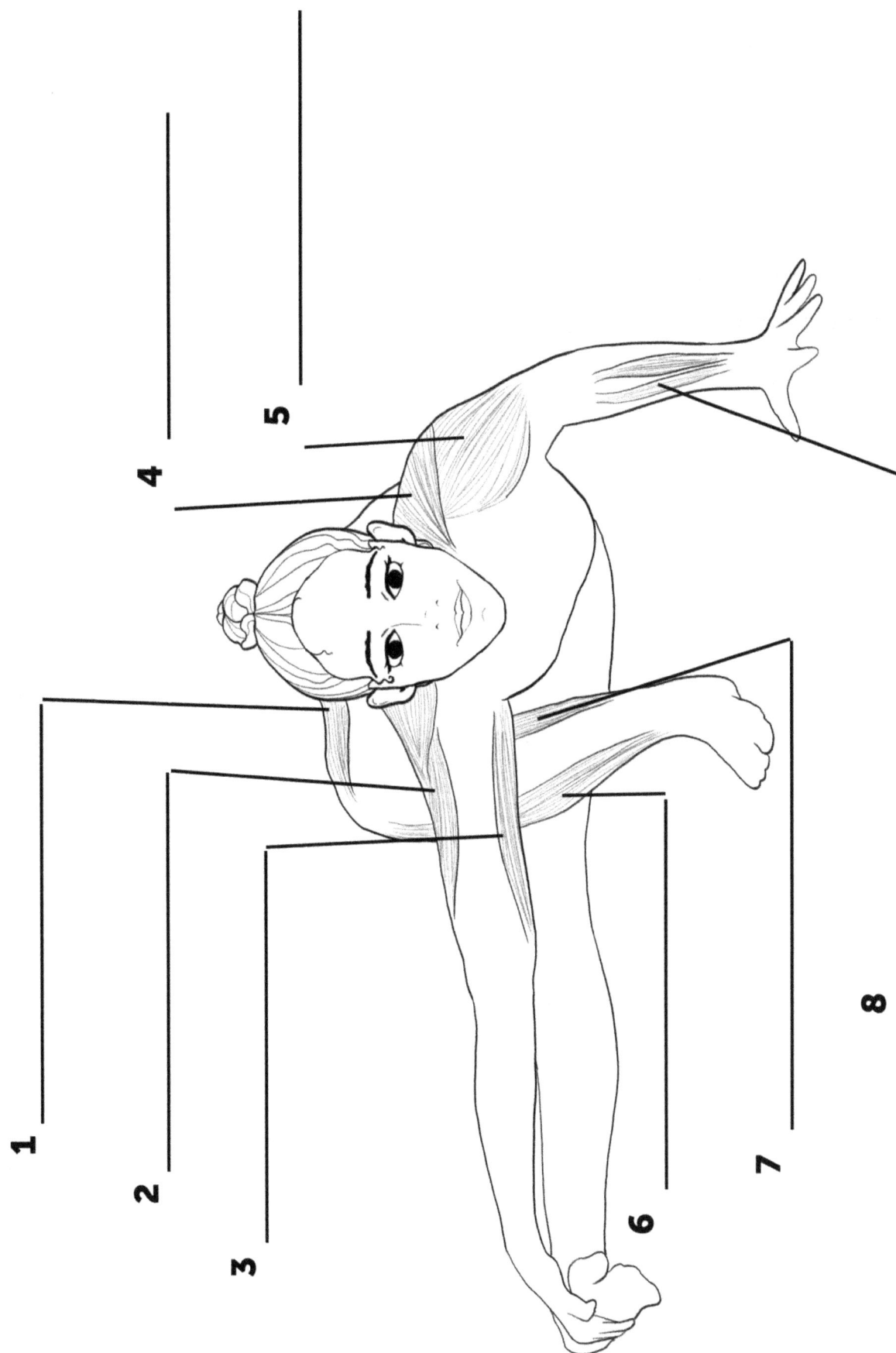

1

2

3

4

5

6

7

8

13. MASTER BABY GRASSHOPPER POSE

1. QUADRICEPS

2. TRICEPS BRACHII

3. BICEPS BRACHII

4. TRAPEZIUS

5. DELTOID

6. TIBIALIS ANTERIOR

7. GASTROCNEMIUS

8. PRONATORS

14. UPWARD-FACING TWO-FOOTED STAFF POSE

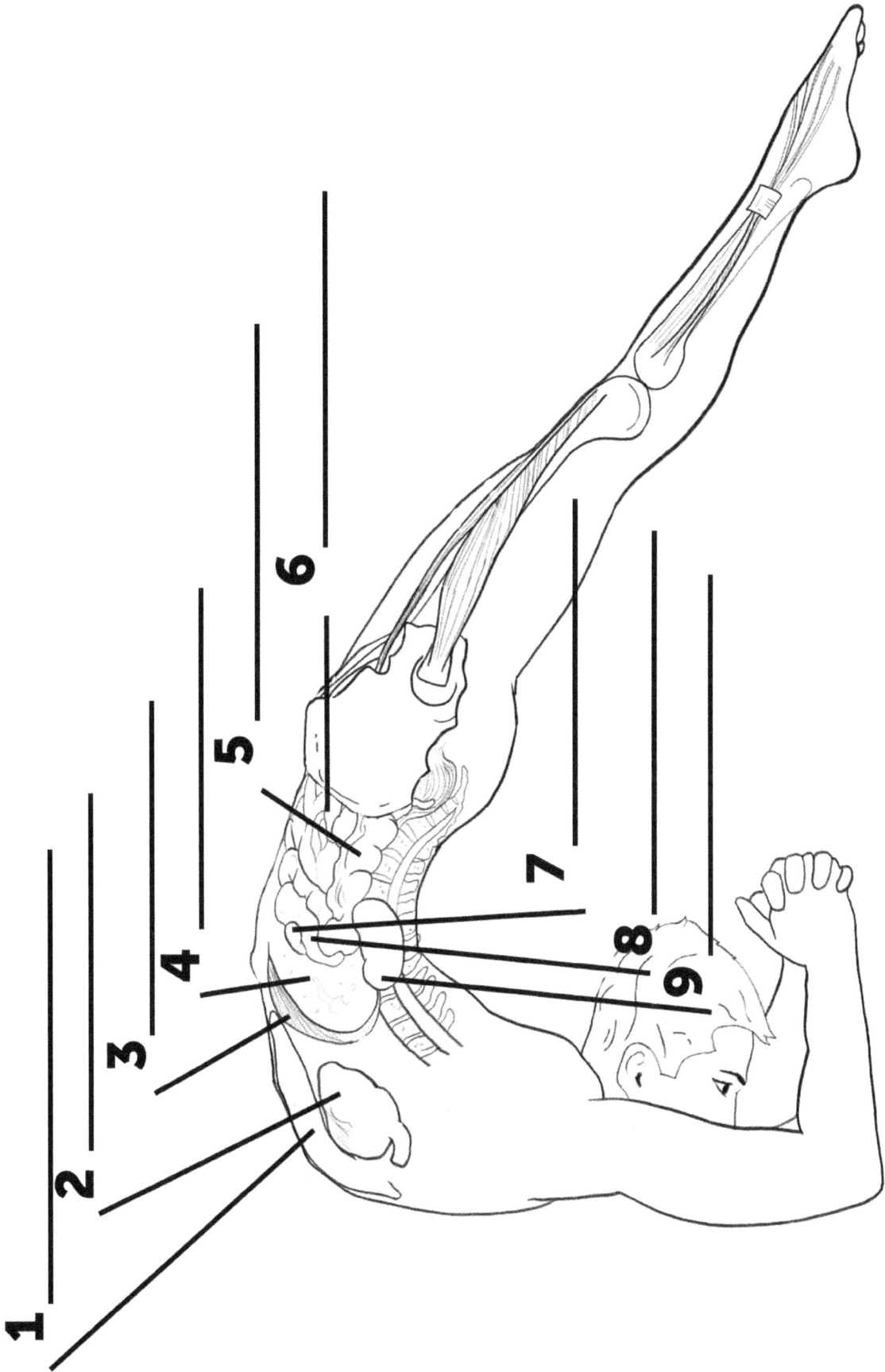

14. UPWARD-FACING TWO-FOOTED STAFF POSE

1.   LUNGS

2.   HEART

3.   DIAPHRAGM

4.   LIVER

5.   ASCENDING COLON

6.   COILS OF SMALL INTESTINE

7.   GALLBLADDER

8.   STOMACH

9.   KIDNEY

15. BHARADVAJA'S TWIST

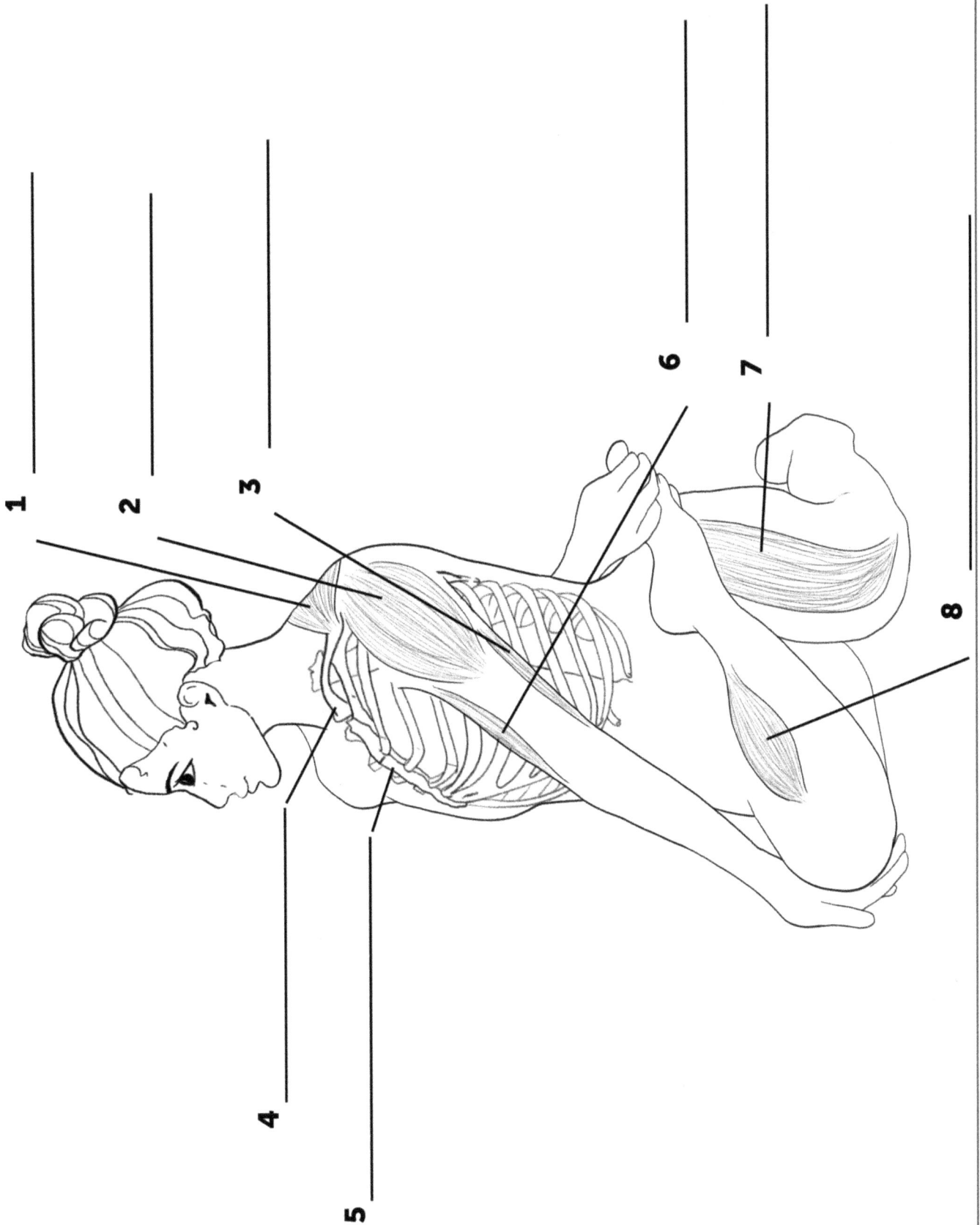

1

2

3

4

5

6

7

8

15. BHARADVAJA'S TWIST

1. TRAPEZIUS

2. DELTOID

3. TRICEPS BRACHII

4. COLLARBONE

5. STERNUM

6. BICEPS BRACHII

7. QUADRICEPS

8. GASTROCNEMIUS

16. EIGHT ANGLE POSE

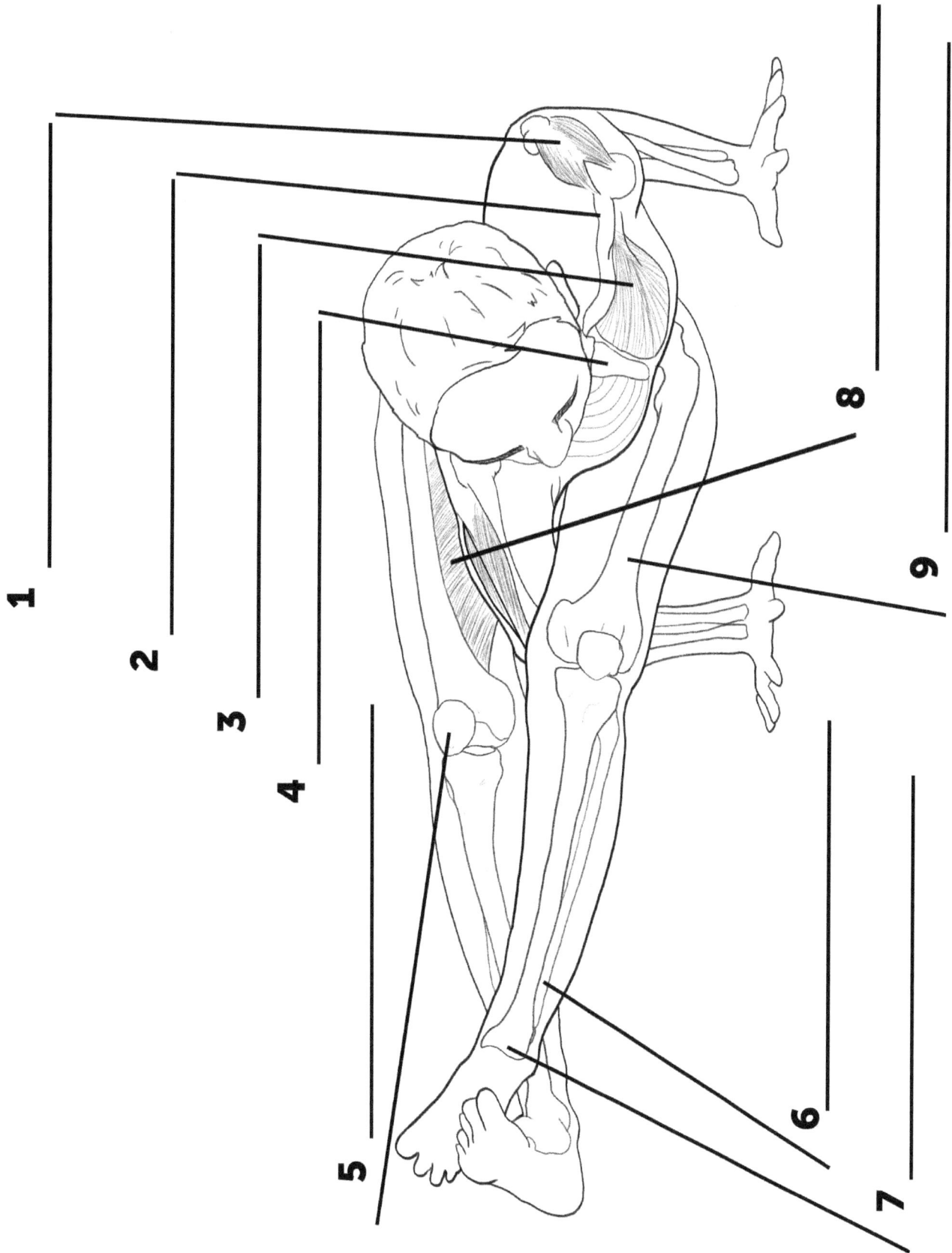

1

2

3

4

5

6

7

8

9

16. EIGHT ANGLE POSE

1. TRICEPS BRACHII
2. COLLARBONE
3. PECTORALIS MAJOR
4. STERNUM
5. PATELLA
6. FIBULA
7. TIBIA
8. ADDUCTORS
9. FEMUR

17. SAGE HALF BOUND LOTUS POSE

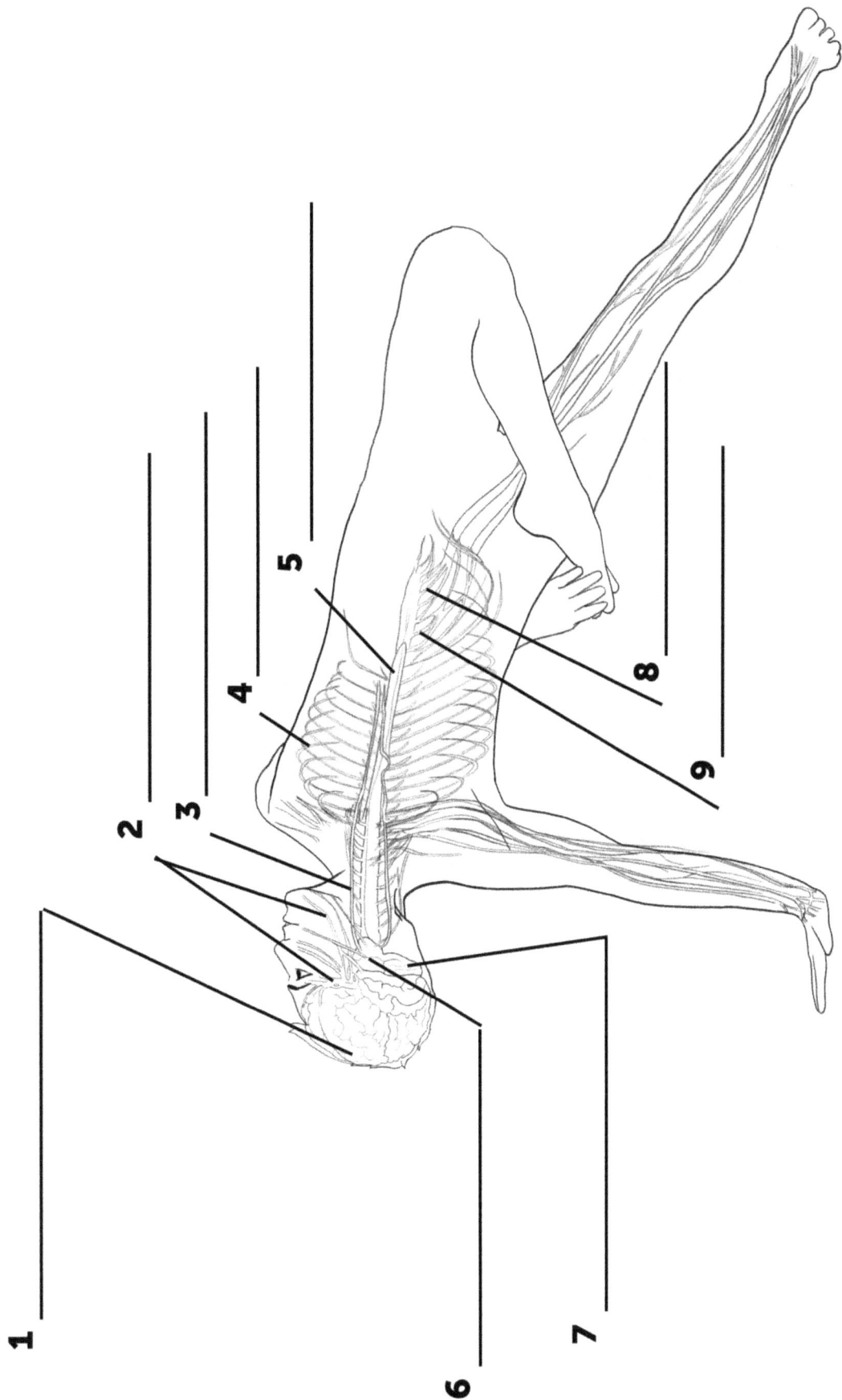

1

2

3

4

5

6

7

8

9

17. SAGE HALF BOUND LOTUS POSE

1. CEREBRUM
2. CRANIAL NERVES
3. VAGUS
4. INTERCOSTALS
5. SPINAL CORD
6. BRAINSTEM
7. CEREBELLUM
8. SACRAL PLEXUS
9. LUMBAR PLEXUS

18. SHOULDER PRESSING POSE

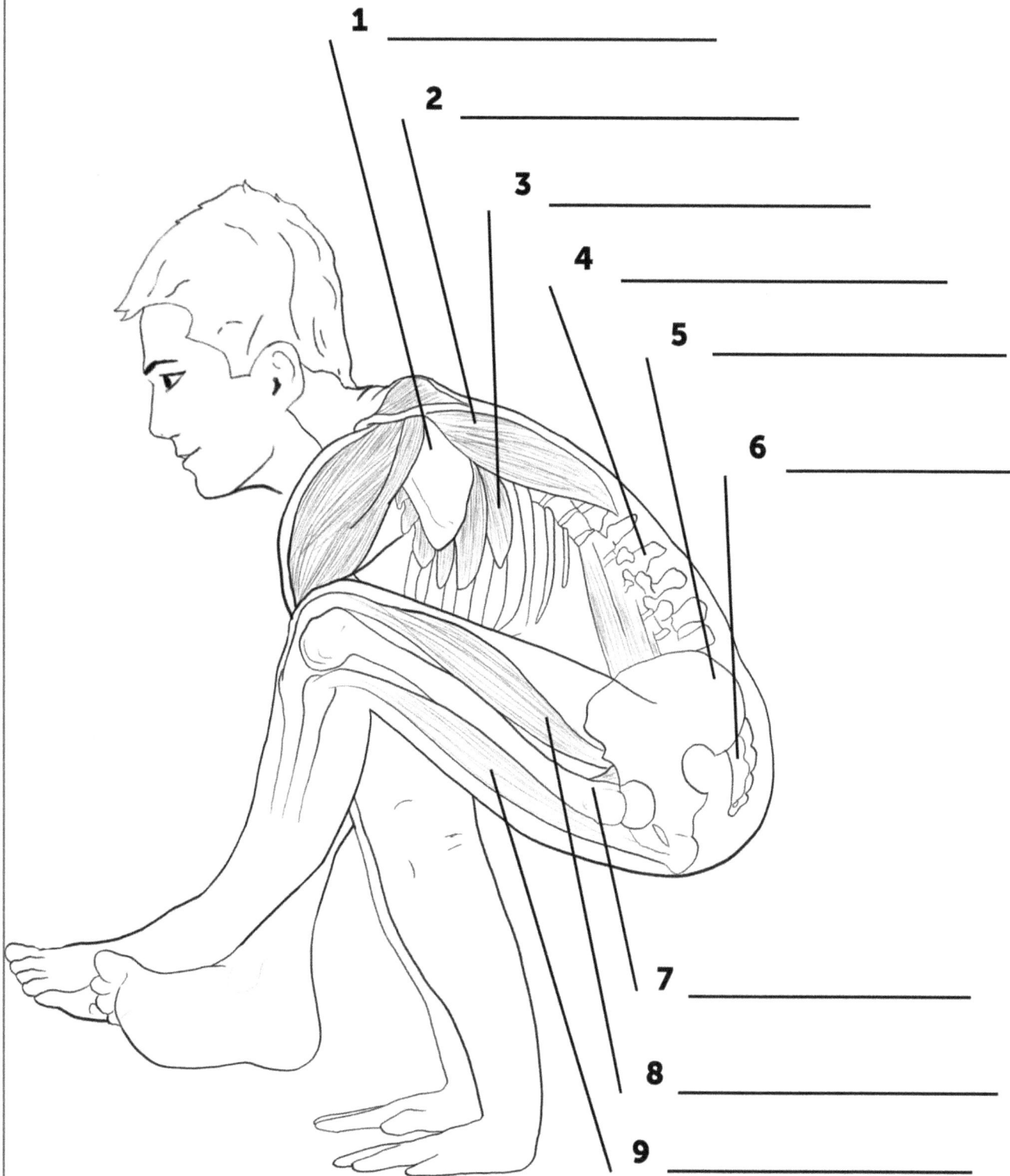

1 _____

2 _____

3 _____

4 _____

5 _____

6 _____

7 _____

8 _____

9 _____

18. SHOULDER PRESSING POSE

1. SCAPULA

2. RHOMBOIDS

3. SERRATUS ANTERIOR

4. SPINE

5. PELVIS

6. SACRUM

7. FEMUR

8. QUADRICEPS

9. HAMSTRINGS

19. SUPER SOLDIER

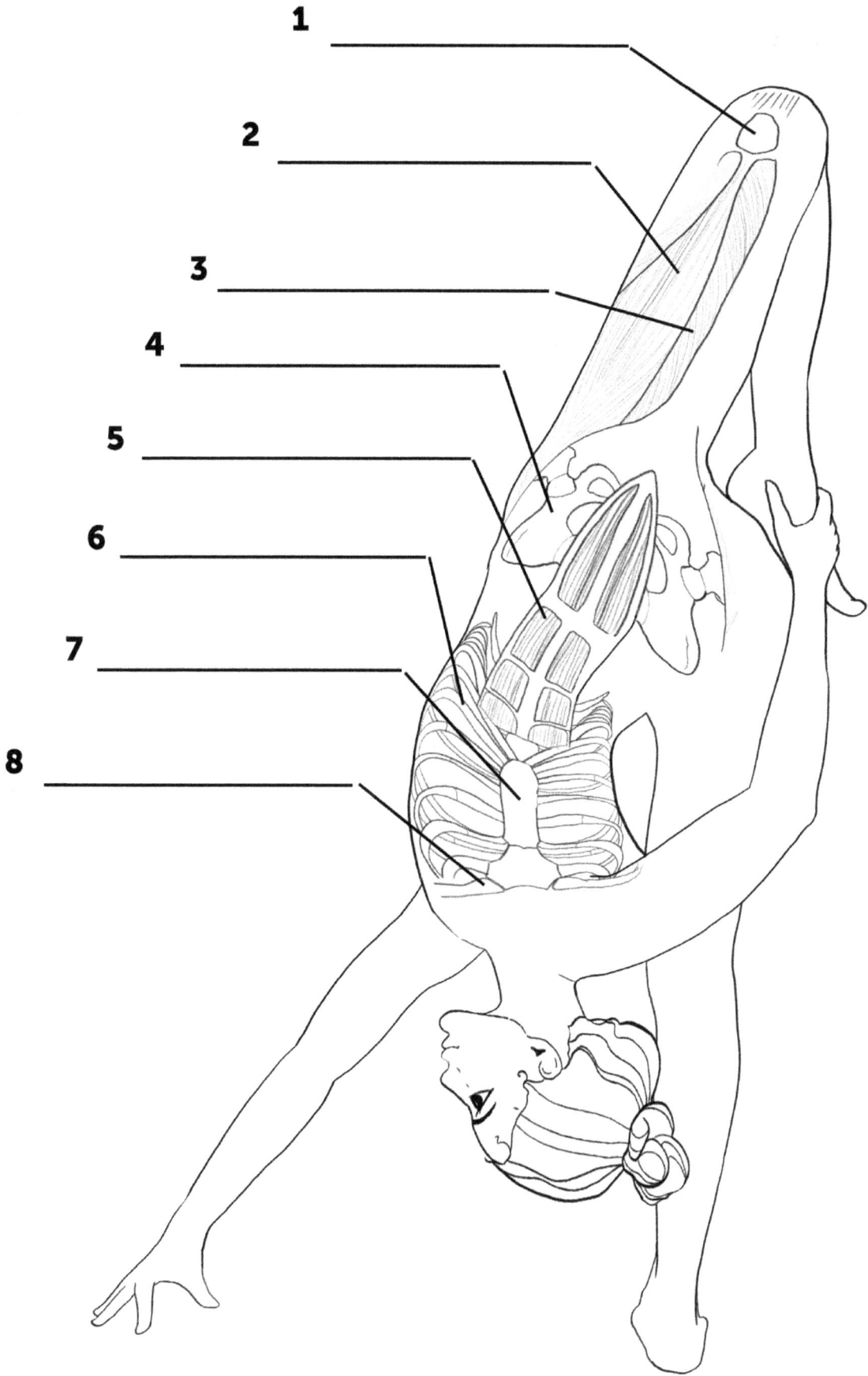

1 _____

2 _____

3 _____

4 _____

5 _____

6 _____

7 _____

8 _____

19. SUPER SOLDIER

1. PATELLA
2. RECTUS FEMORIS
3. VASTUS MEDIALIS
4. PELVIS
5. RECTUS ABDOMINIS
6. RIBS
7. STERNUM
8. COLLARBONE

20. MONKEY POSE

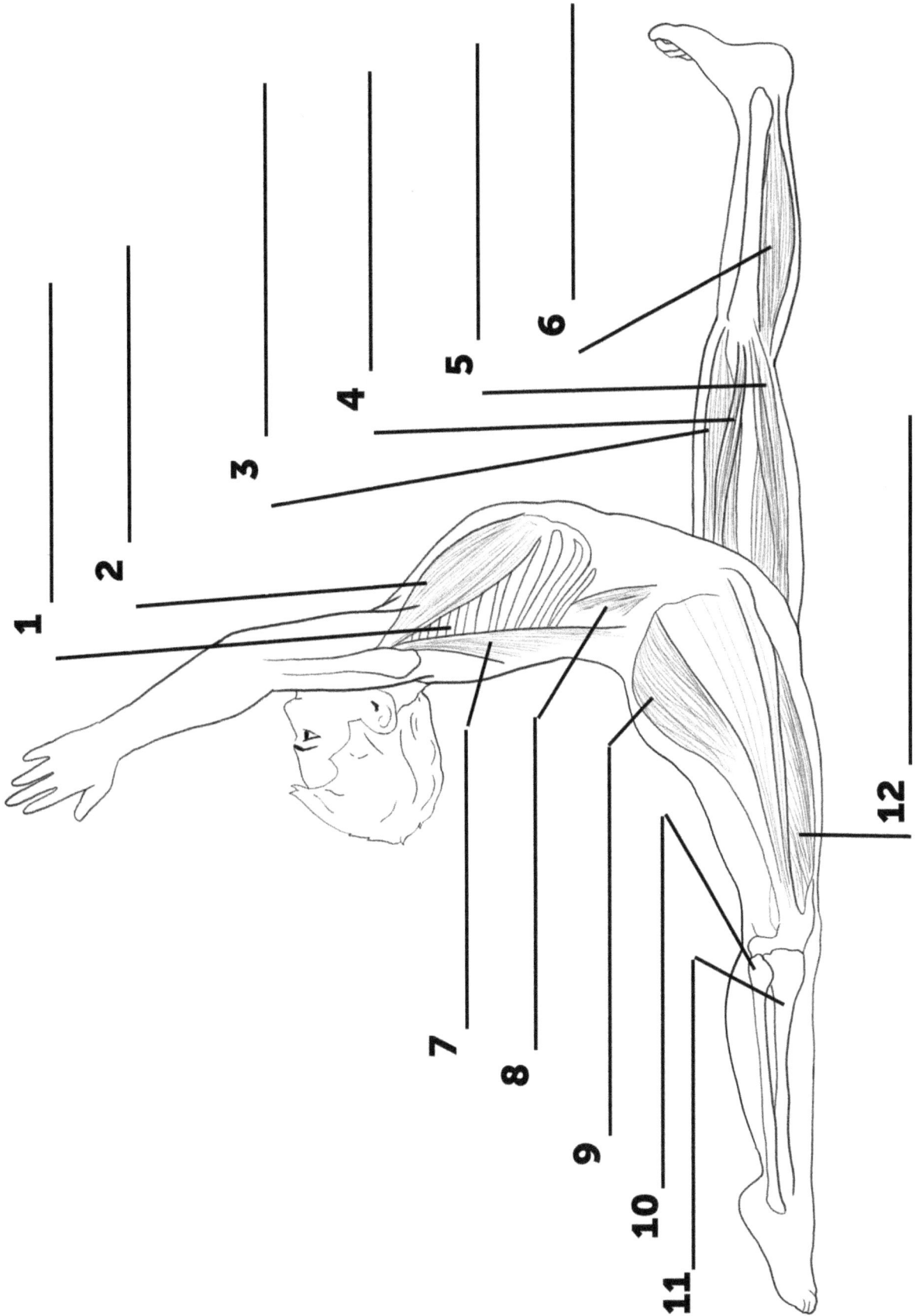

1

2

3

4

5

6

7

8

9

10

11

12

20. MONKEY POSE

1. RIBS
2. PECTORALIS MAJOR
3. RECTUS FEMORIS
4. SARTORIUS
5. HAMSTRINGS
6. GASTROCNEMIUS
7. LATISSIMUS DORSI
8. ERECTOR SPINAE
9. GLUTEUS MAXIMUS
10. FIBULA
11. TIBIA
12. QUADRICEPS

21. SEATED WIDE ANGLE POSE

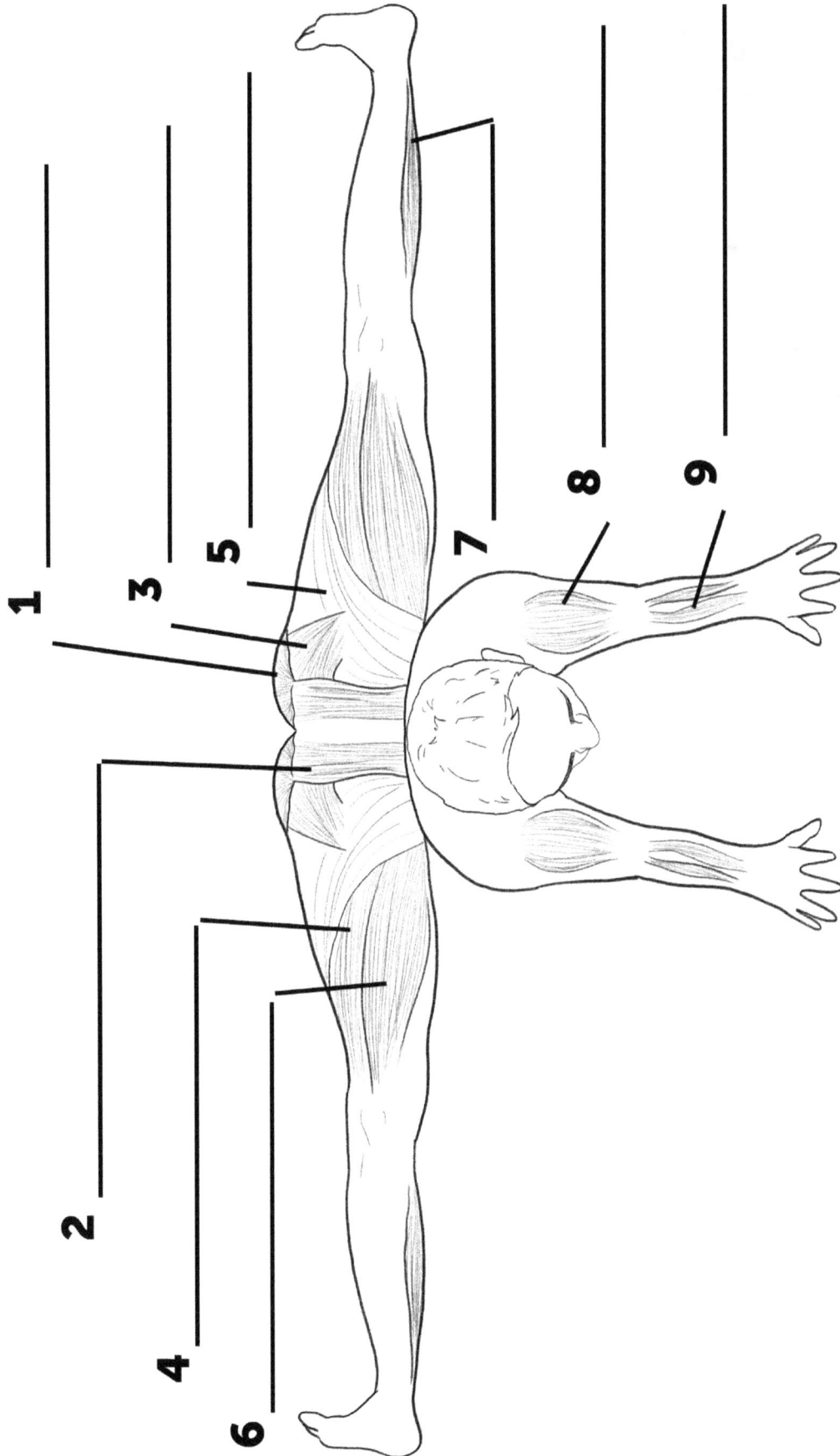

1

2

3

4

5

6

7

8

9

21. SEATED WIDE ANGLE POSE

1.   GLUTEUS MAXIMUS

2.   ERECTOR SPINAE

3.   GLUTEUS MEDIUS

4.   VASTUS LATERALIS

5.   ILIOTIBIAL BAND

6.   RECTUS FEMORIS

7.   GASTROCNEMIUS

8.   DELTOID

9.   PRONATORS

22. EXTENDED BALANCING LIZARD

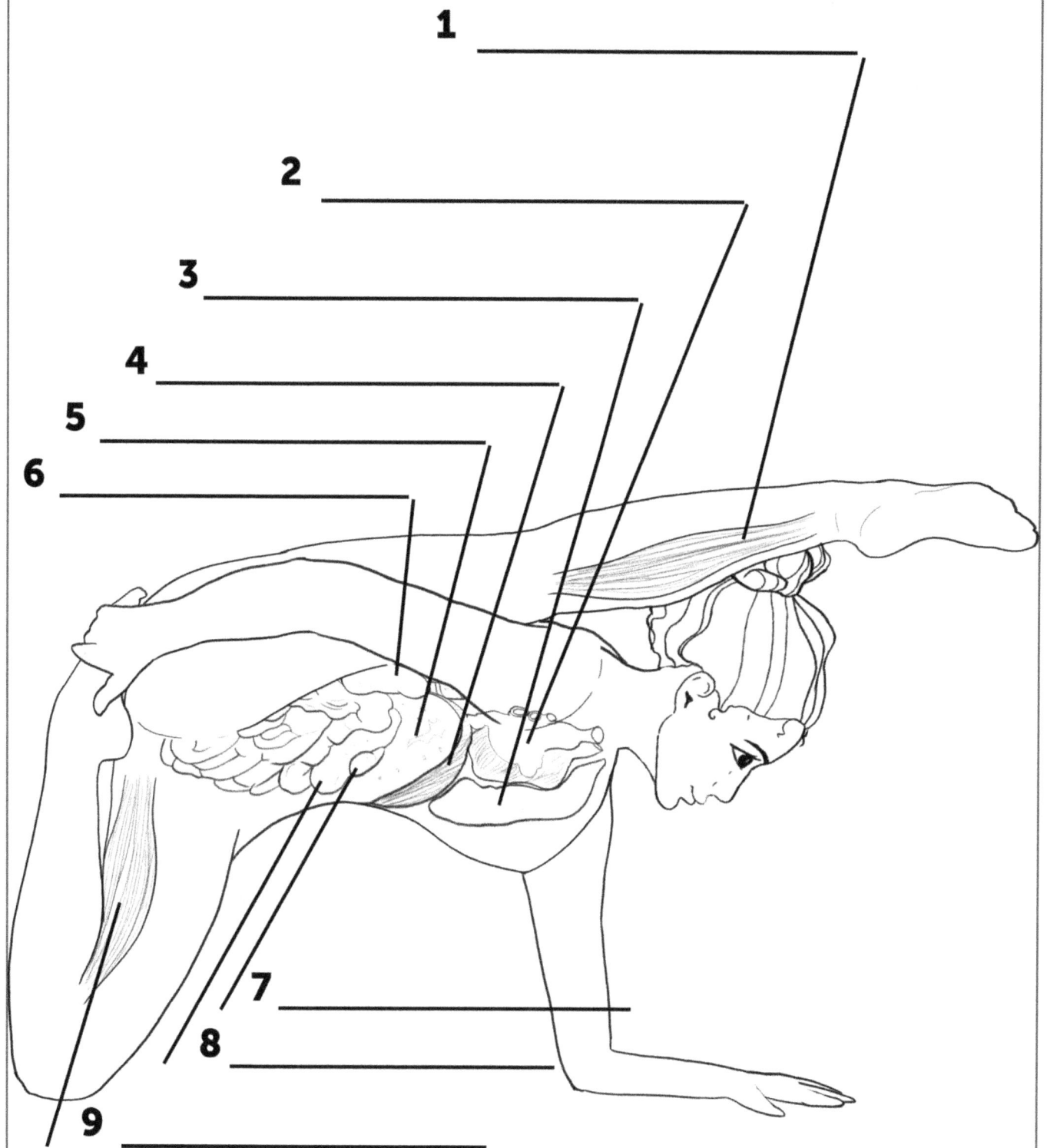

1 _____

2 _____

3 _____

4 _____

5 _____

6 _____

7 _____

8 _____

9 _____

22. EXTENDED BALANCING LIZARD

1. GASTROCNEMIUS
2. HEART
3. LUNGS
4. DIAPHRAGM
5. LIVER
6. KIDNEY
7. GALLBLADDER
8. STOMACH
9. HAMSTRINGS

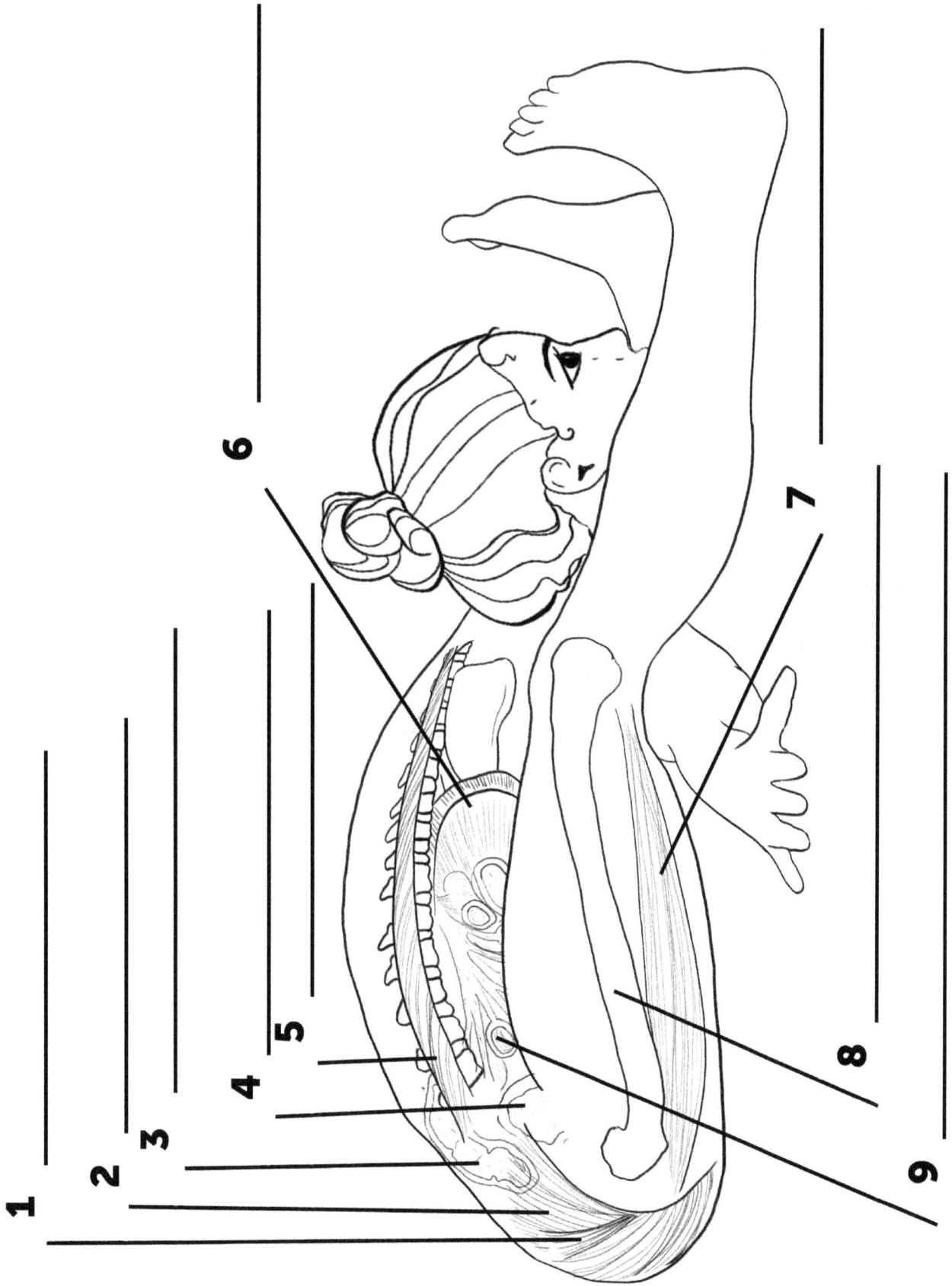

23. KURMASANA

1

2

3

4

5

6

7

8

9

23. KURMASANA

1. PIRIFORMIS
2. GLUTEUS MAXIMUS
3. RECTUM
4. URINARY BLADDER
5. SPINAL MUSCLES
6. DIAPHRAGM
7. HAMSTRINGS
8. FEMUR
9. COILS OF SMALL INTESTINE

24. VIPARITA SALABHASANA

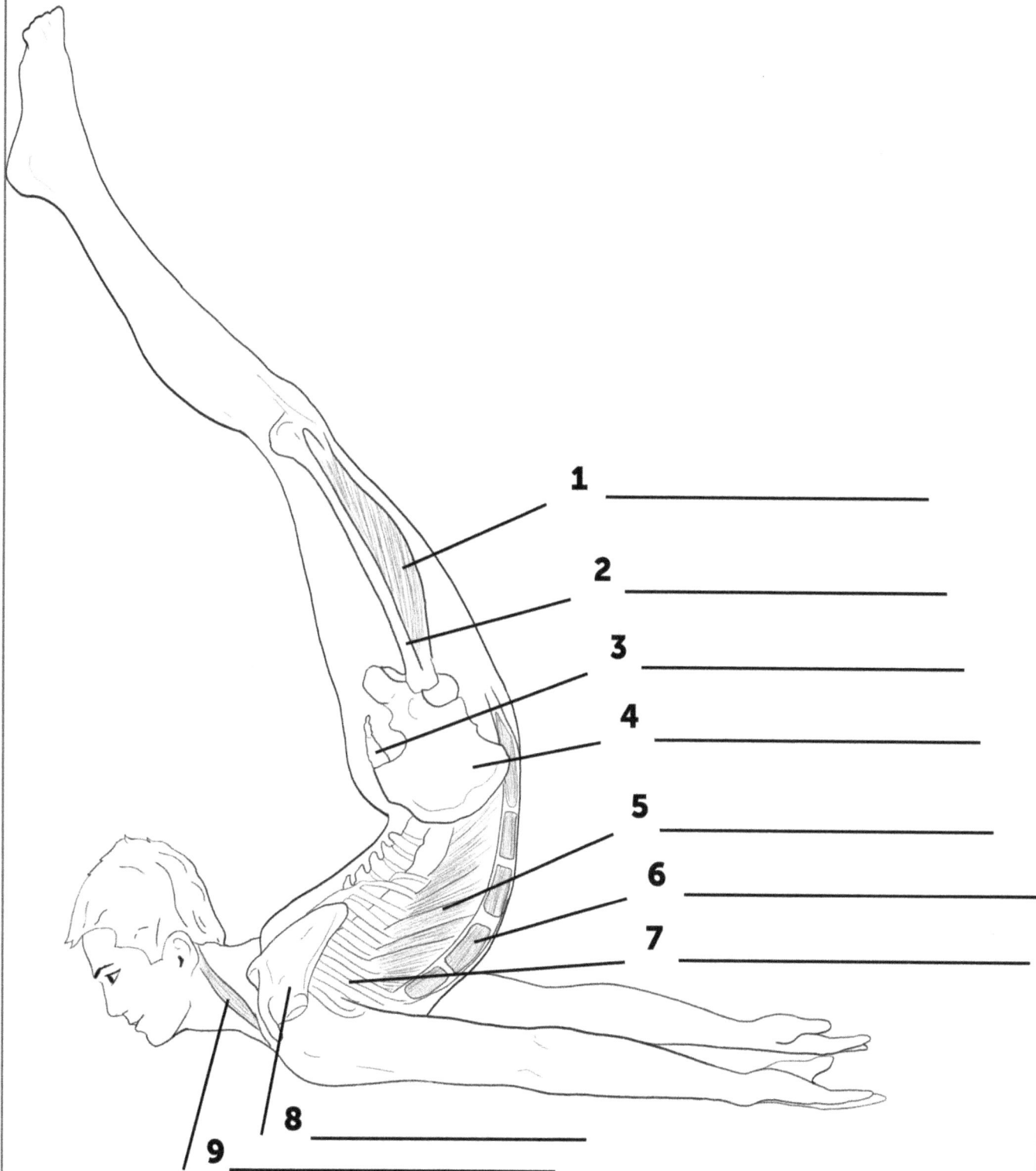

1 _____

2 _____

3 _____

4 _____

5 _____

6 _____

7 _____

8 _____

9 _____

24. VIPARITA SALABHASANA

1. QUADRICEPS
2. FEMUR
3. SACRUM
4. PELVIS
5. EXTERNAL OBLIQUE
6. RECTUS ABDOMINIS
7. RIBS
8. SCAPULA
9. STERNOCLEIDOMASTOID

25. SLEEPING YOGI POSE

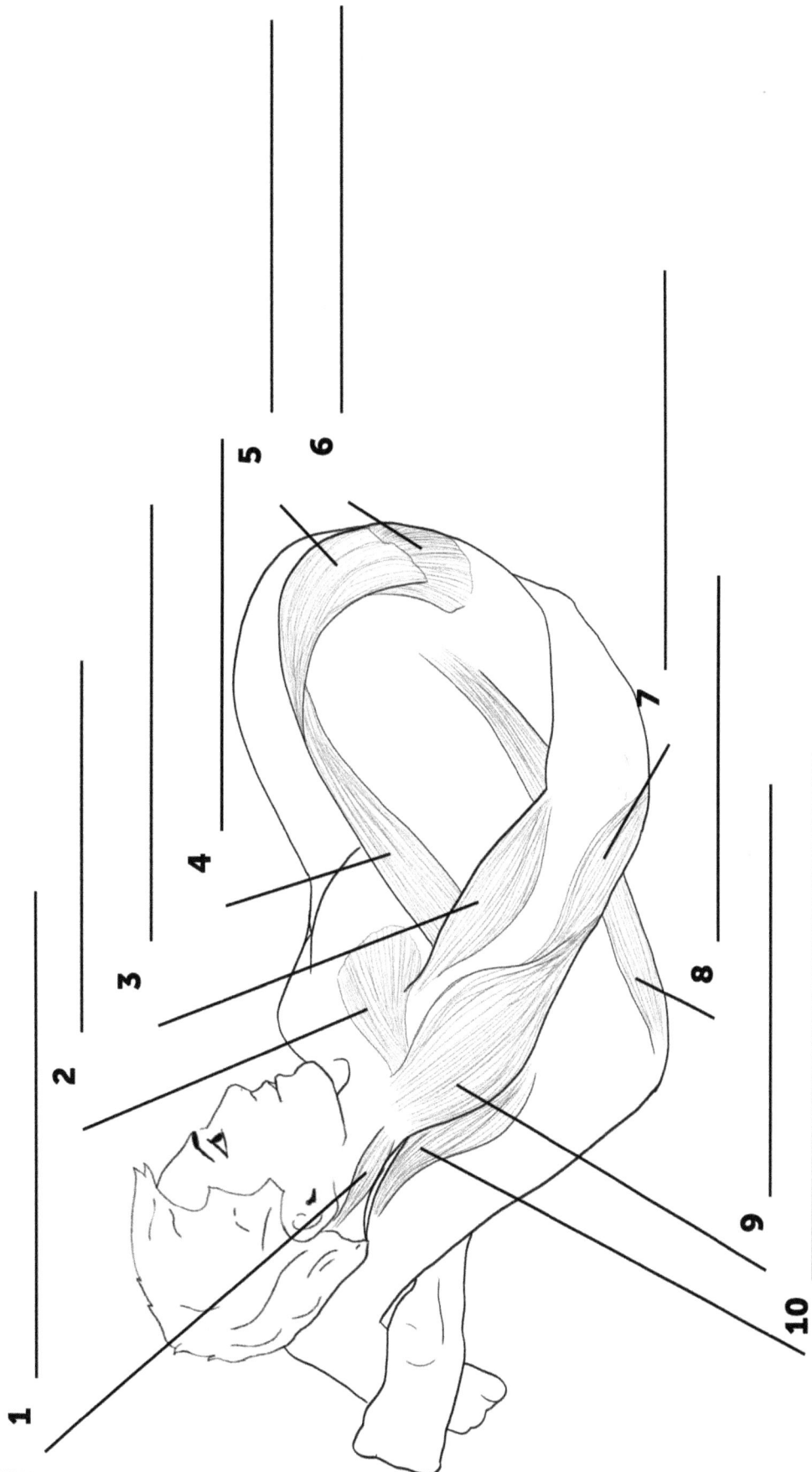

25. SLEEPING YOGI POSE

1. STERNOCLEIDOMASTOID
2. PECTORALIS MAJOR
3. BICEPS BRACHII
4. HAMSTRINGS
5. GLUTEUS MAXIMUS
6. GLUTEUS MEDIUS
7. TRICEPS BRACHII
8. QUADRICEPS
9. DELTOID
10. GASTROCNEMIUS

26. DOVE POSE

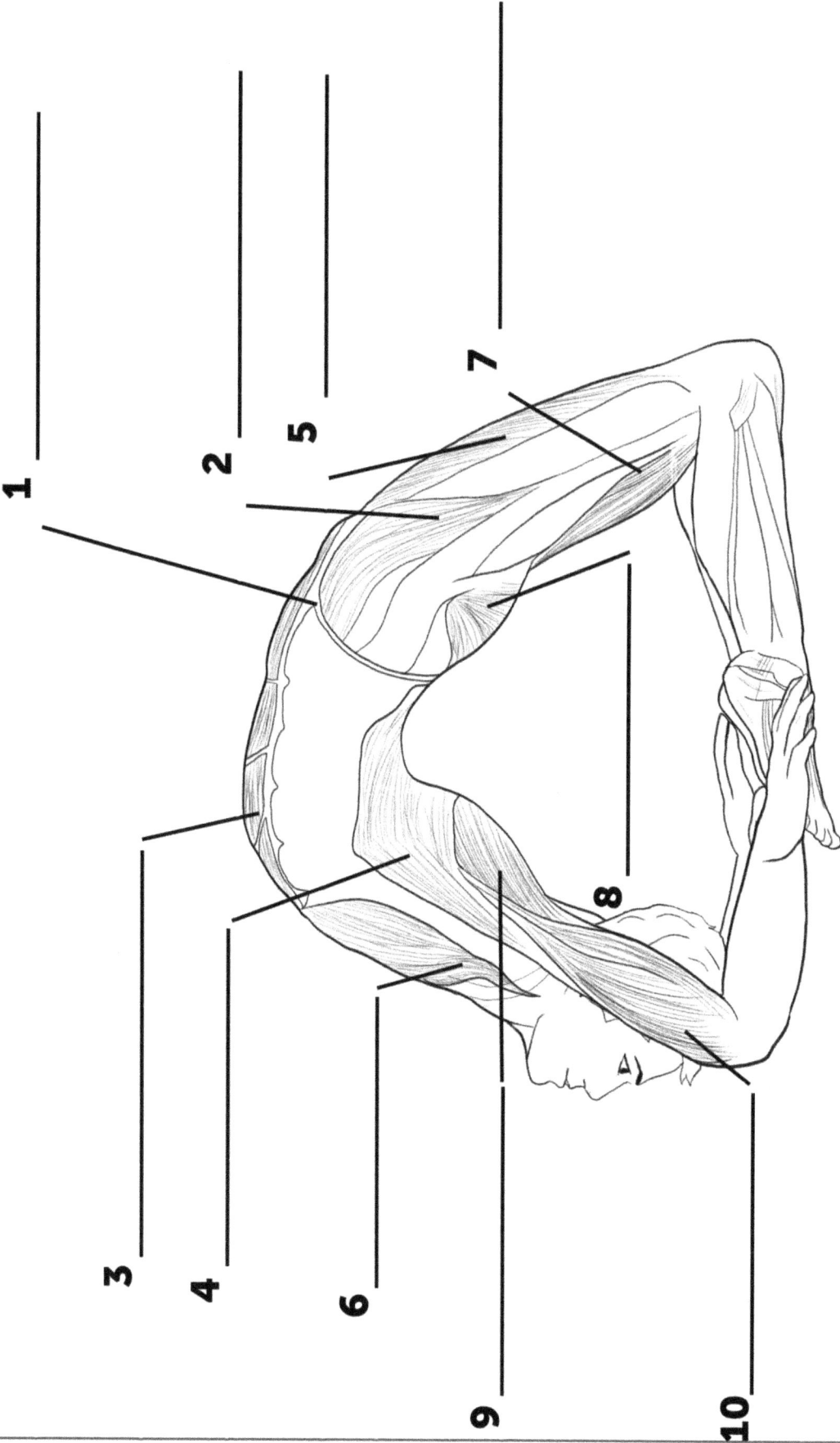

1 _____

2 _____

5 _____

7 _____

3 _____

4 _____

6 _____

8 _____

9 _____

10 _____

26. DOVE POSE

1. ILIOPSOAS
2. TENSOR FASCIA LATA
3. RECTUS ABDOMINIS
4. LATISSIMUS DORSI
5. QUADRICEPS
6. PECTORALIS MAJOR
7. HAMSTRINGS
8. GLUTEUS MAXIMUS
9. ERECTOR SPINAE
10. TRICEPS BRACHII

27. BOUND ANGLE HEADSTAND POSE

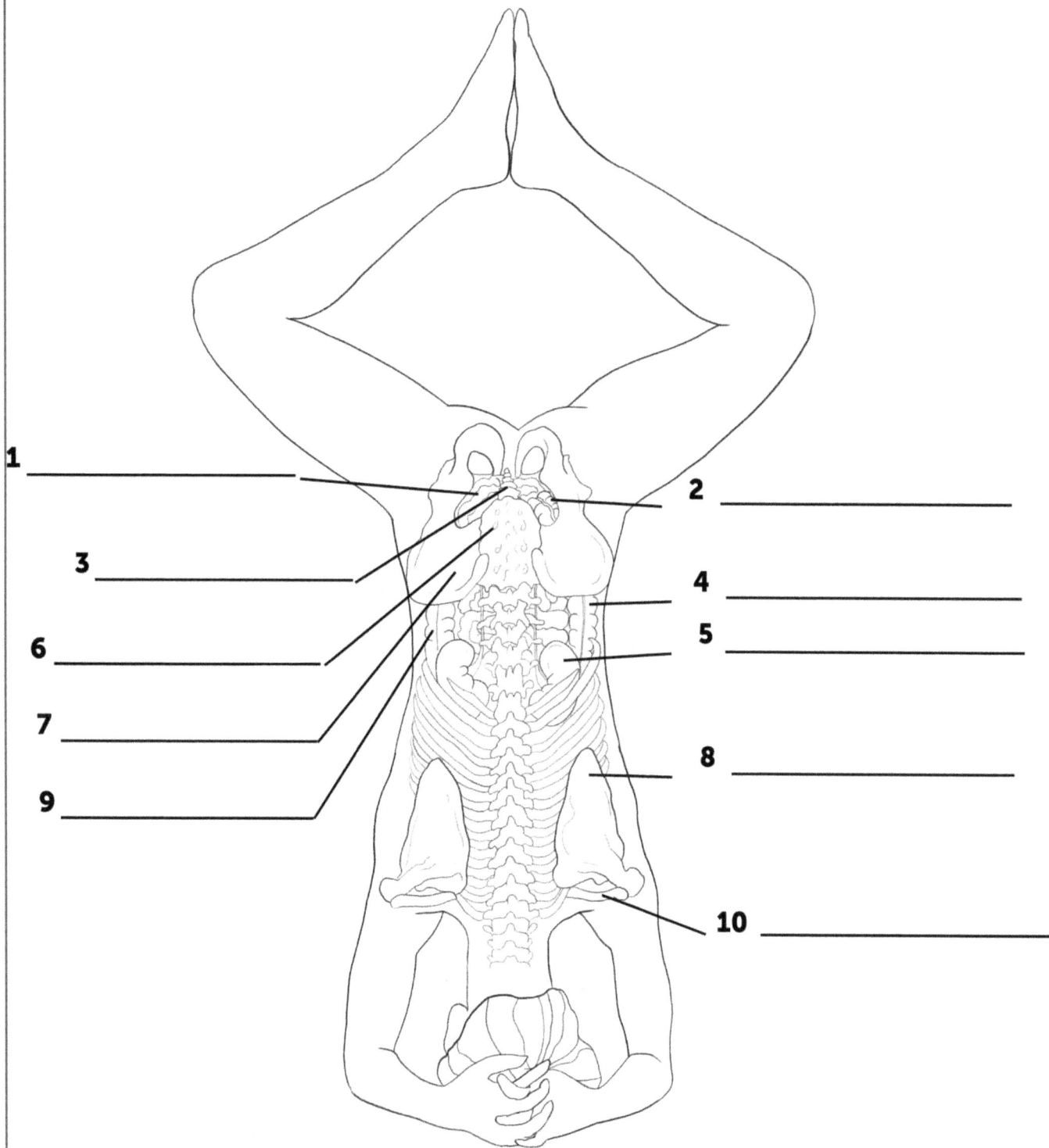

1 _____

2 _____

3 _____

4 _____

5 _____

6 _____

7 _____

8 _____

9 _____

10 _____

27. BOUND ANGLE HEADSTAND POSE

1. COILS OF SMALL INTESTINE

2. SIGMOID COLON

3. COCCYX

4. DESCENDING COLON

5. KIDNEY

6. SACRUM

7. PELVIS

8. SCAPULA

9. ASCENDING COLON

10. COLLARBONE

28. VISVAMITRASANA II

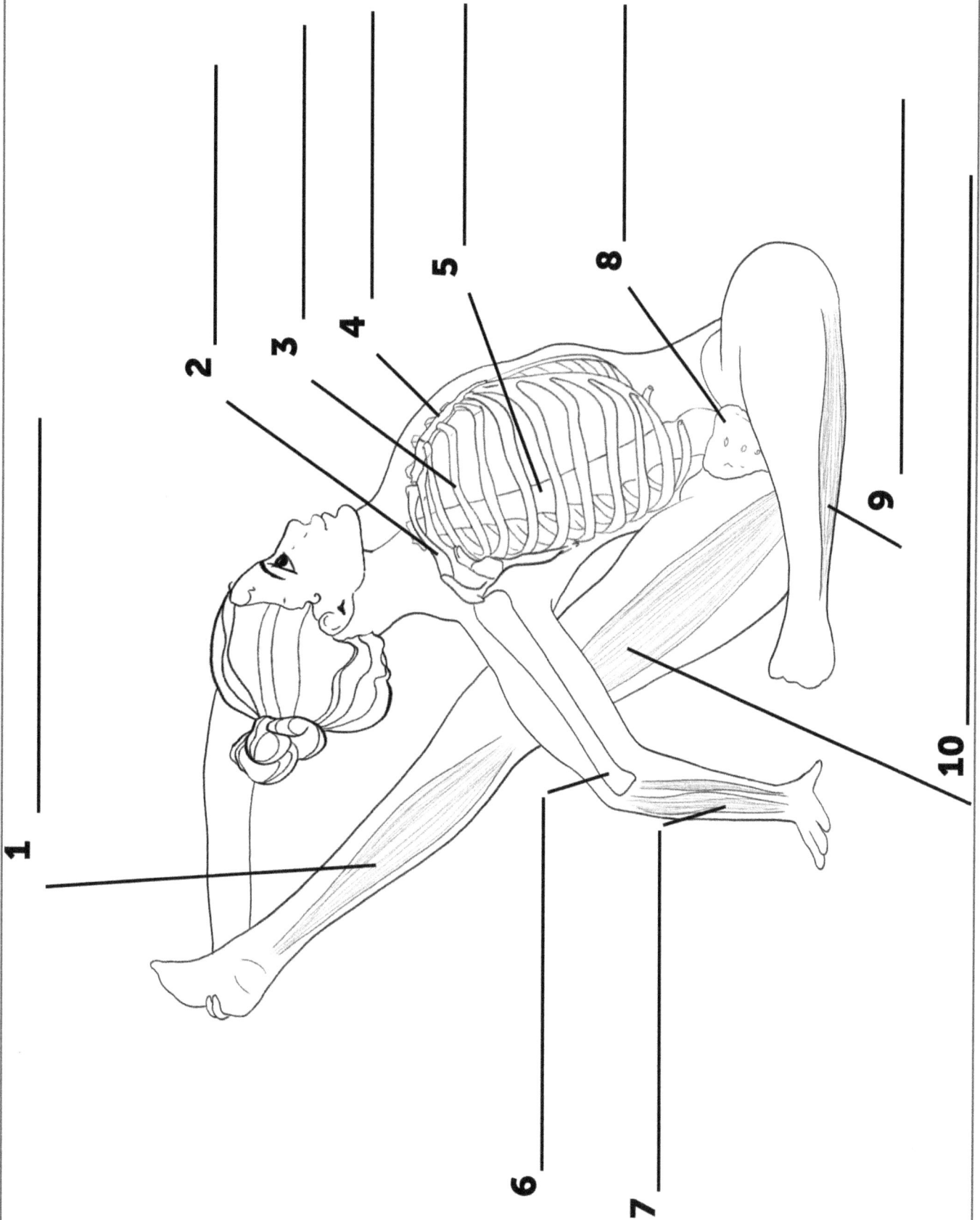

28. VISVAMITRASANA II

1. GASTROCNEMIUS
2. COLLARBONE
3. RIBS
4. STERNUM
5. SPINE
6. HUMERUS
7. PRONATORS
8. SACRUM
9. TIBIALIS ANTERIOR
10. HAMSTRINGS

29. LOTUS IN SHOULDER STAND POSE

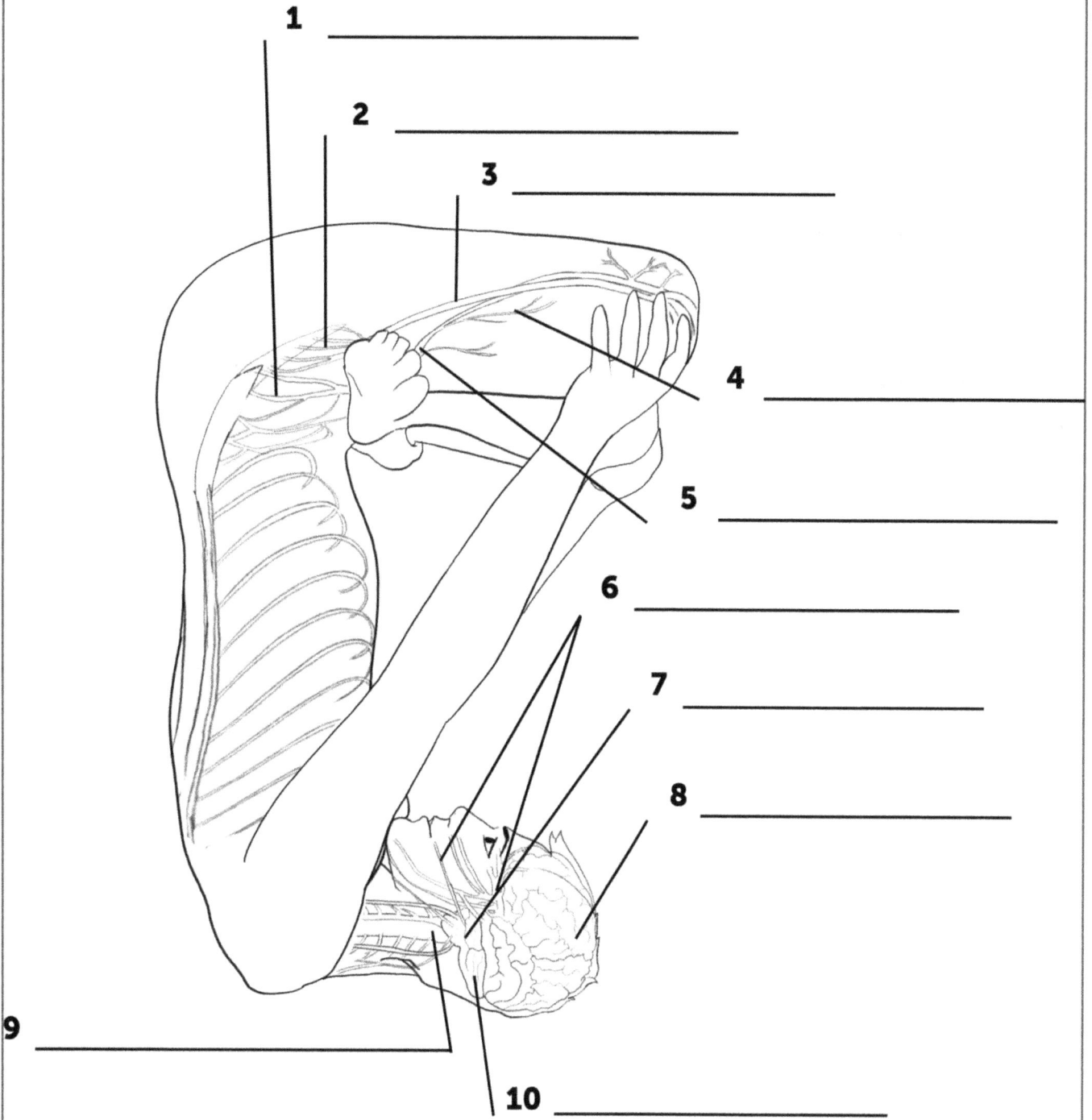

1 _____

2 _____

3 _____

4 _____

5 _____

6 _____

7 _____

8 _____

9 _____

10 _____

29. LOTUS IN SHOULDER STAND POSE

1. LUMBAR PLEXUS

2. SACRAL PLEXUS

3. SCIATIC

4. MUSCULAR BRANCHES OF FEMORAL

5. FEMORAL

6. CRANIAL NERVES

7. BRAINSTEM

8. CEREBRUM

9. SPINAL CORD

10. CEREBELLUM

30. ONE LEGGED WHEEL POSE

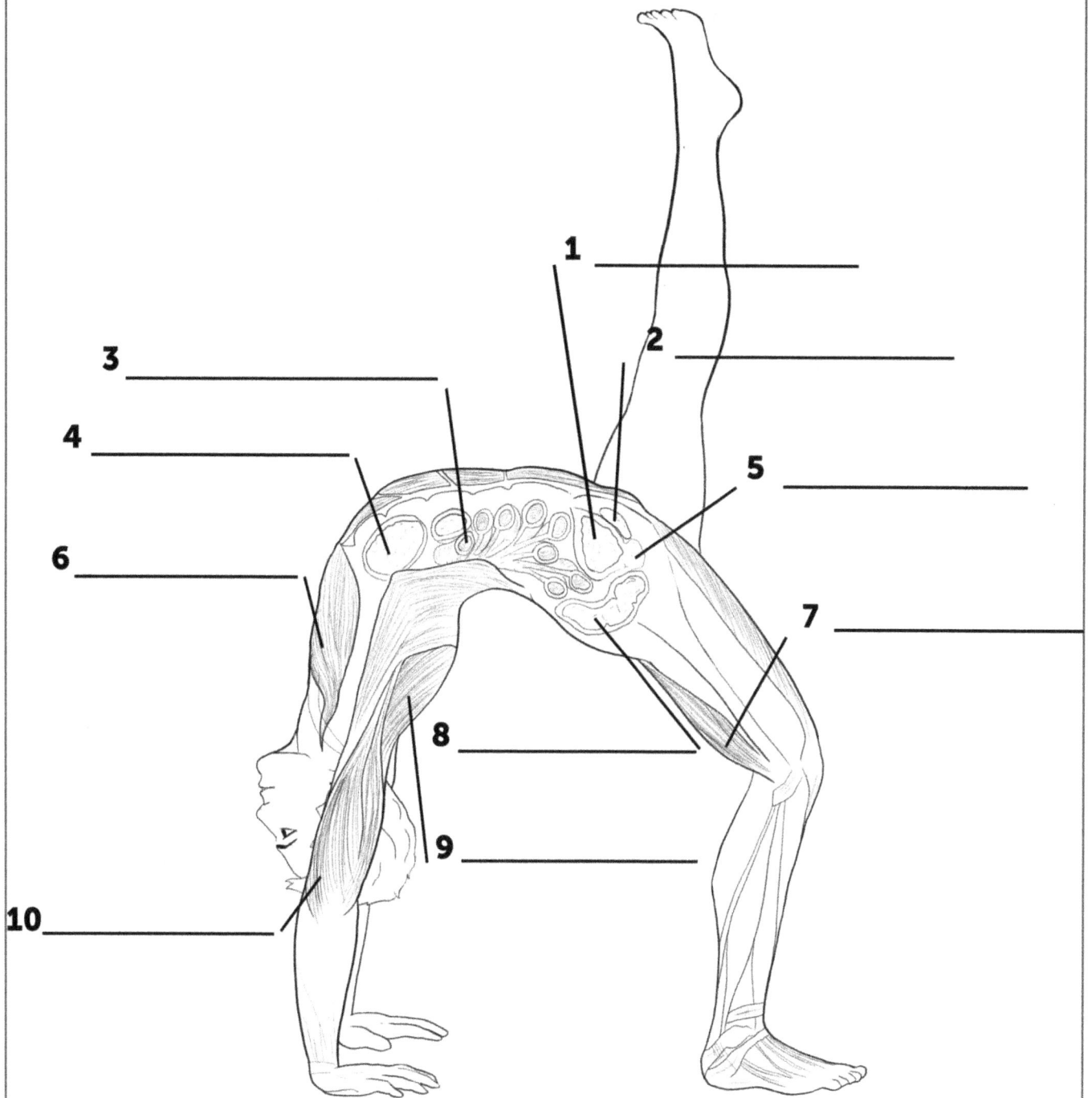

1 _____

2 _____

3 _____

4 _____

5 _____

6 _____

7 _____

8 _____

9 _____

10 _____

30. ONE LEGGED WHEEL POSE

1. URINARY BLADDER

2. PUBIC BONE

3. COILS OF SMALL INTESTINE

4. STOMACH

5. PROSTATE

6. PECTORALIS MAJOR

7. HAMSTRINGS

8. RECTUM

9. ERECTOR SPINAE

10. TRICEPS BRACHII

31. ONE LEGGED HEADSTAND

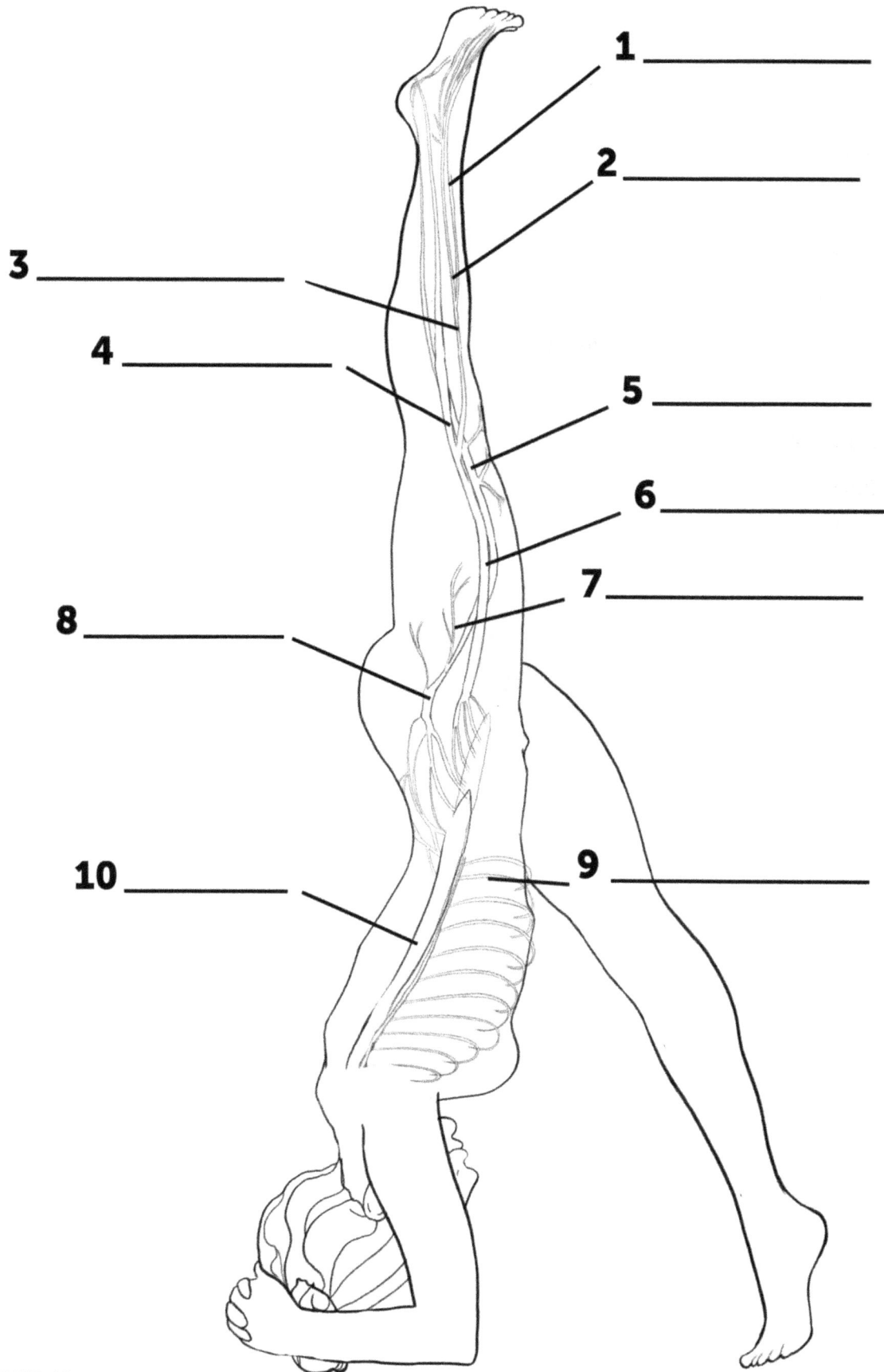

1 _____

2 _____

3 _____

4 _____

5 _____

6 _____

7 _____

8 _____

9 _____

10 _____

31. ONE LEGGED HEADSTAND

1. SUPERFICIAL PERONEAL

2. DEEP PERONEAL

3. COMMON PERONEAL

4. TIBIAL

5. SAPHENOUS

6. SCIATIC

7. MUSCULAR BRANCHES OF FEMORAL

8. FEMORAL

9. INTERCOSTALS

10. SPINAL CORD

32. SUPTA VISVAMITRASANA

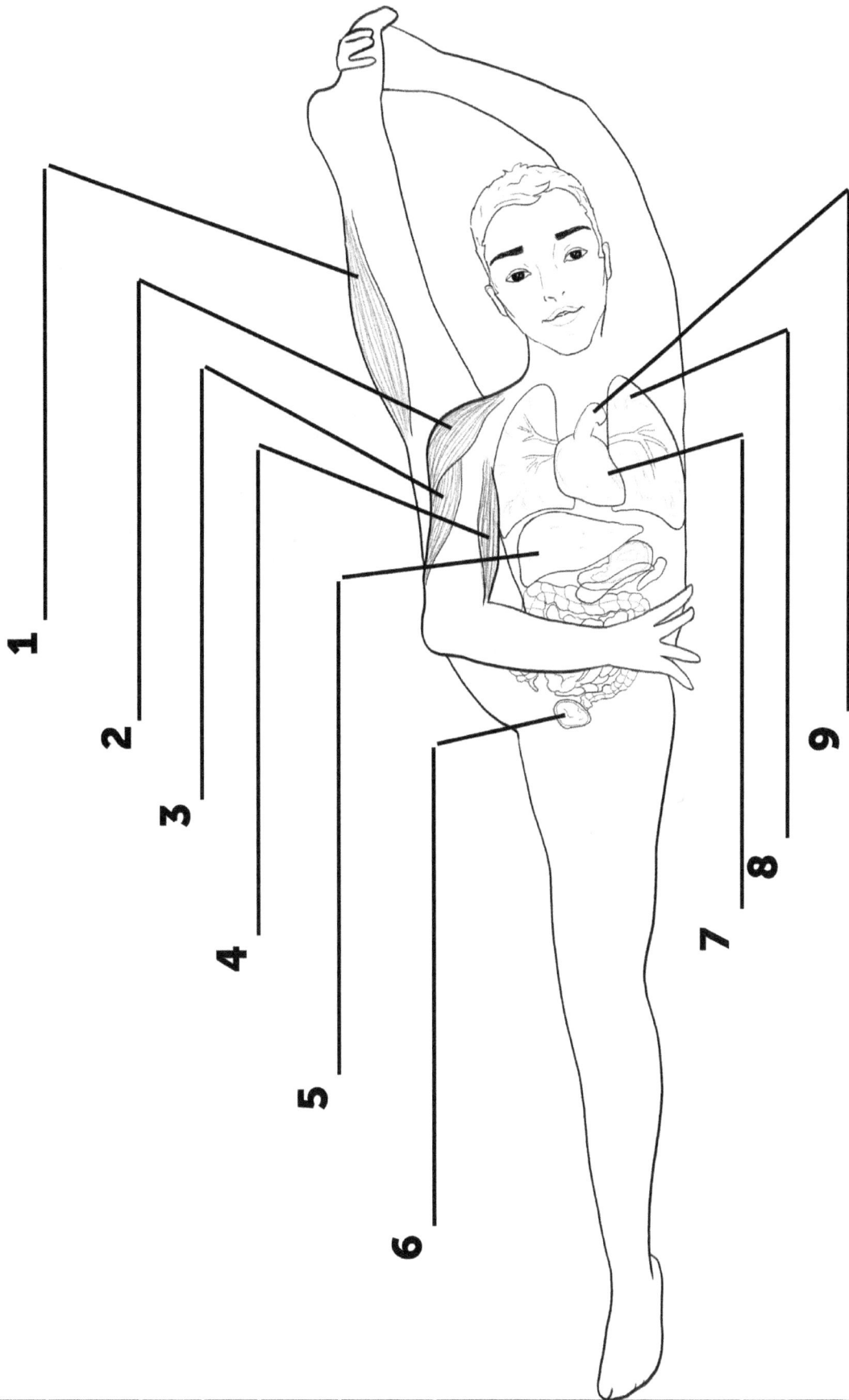

1

2

3

4

5

6

7

8

9

32. SUPTA VISVAMITRASANA

1. GASTROCNEMIUS
2. DELTOID
3. TRICEPS BRACHII
4. BICEPS BRACHII
5. LIVER
6. URINARY BLADDER
7. HEART
8. LUNGS
9. AORTA

33. UPWARD FACING FORWARD BEND POSE

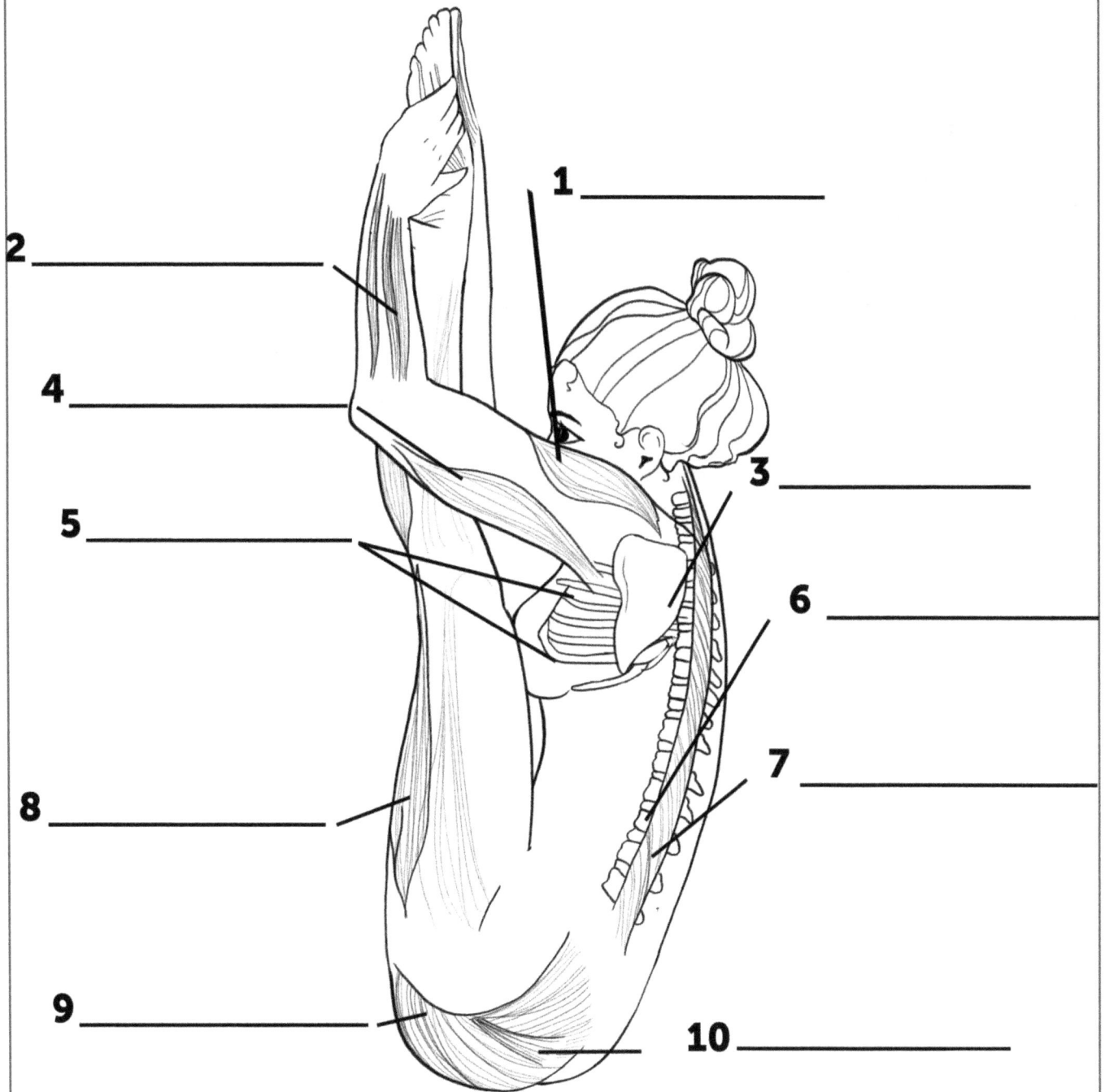

1 _____

2 _____

4 _____

3 _____

5 _____

6 _____

7 _____

8 _____

9 _____

10 _____

33. UPWARD FACING FORWARD BEND POSE

1. DELTOID
2. PRONATORS
3. SCAPULA
4. TRICEPS BRACHII
5. RIBS
6. SPINE
7. SPINAL MUSCLES
8. HAMSTRINGS
9. GLUTEUS MAXIMUS
10. PIRIFORMIS

34. UPWARD FACING WIDE-ANGLE SEATED POSE

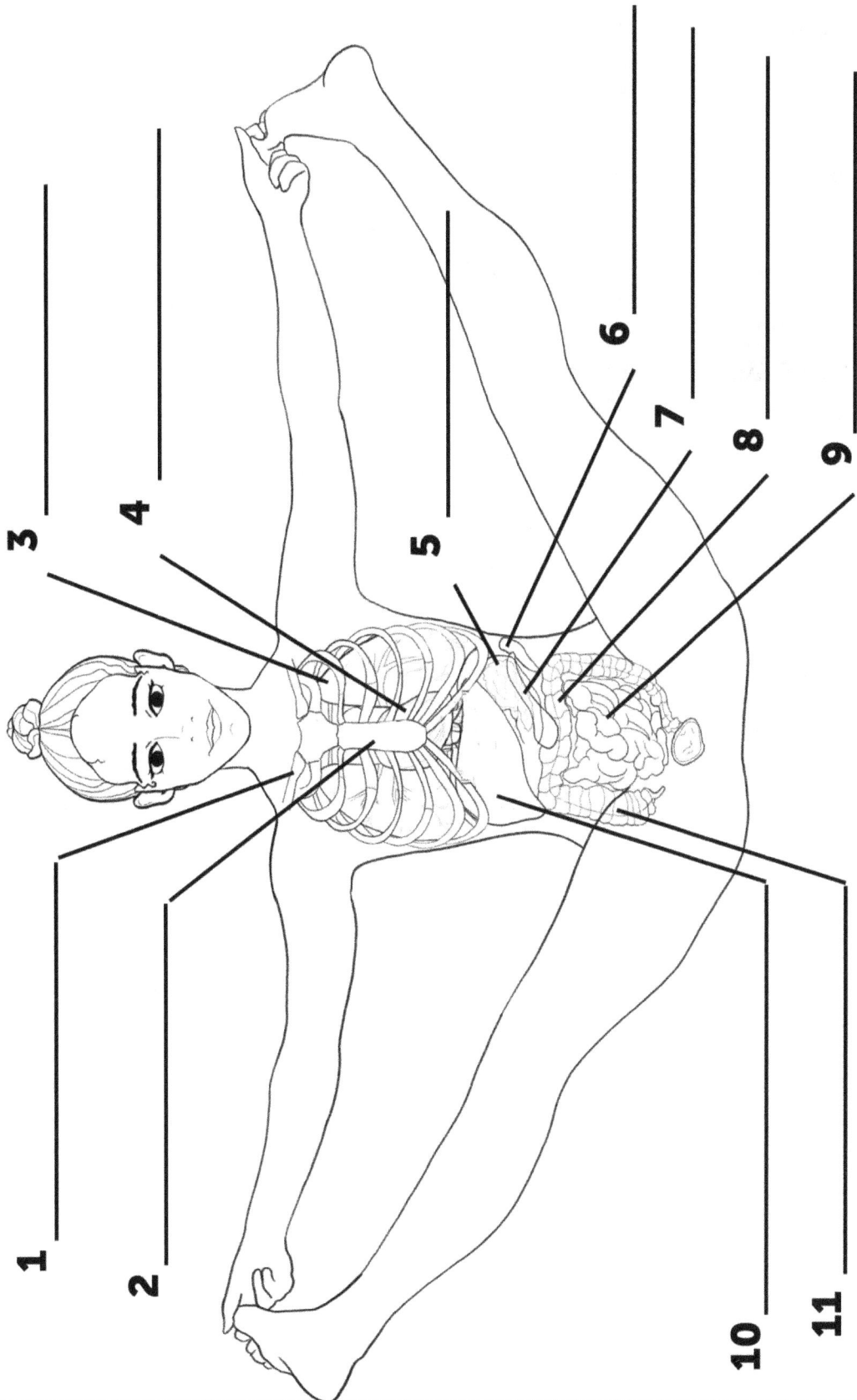

1

2

3

4

5

6

7

8

9

10

11

34. UPWARD FACING WIDE-ANGLE SEATED POSE

1. COLLARBONE
2. STERNUM
3. LUNGS
4. HEART
5. STOMACH
6. SPLEEN
7. PANCREAS
8. TRANSVERSE COLON
9. COILS OF SMALL INTESTINE
10. LIVER
11. ASCENDING COLON

35. VISVAMITRASANA

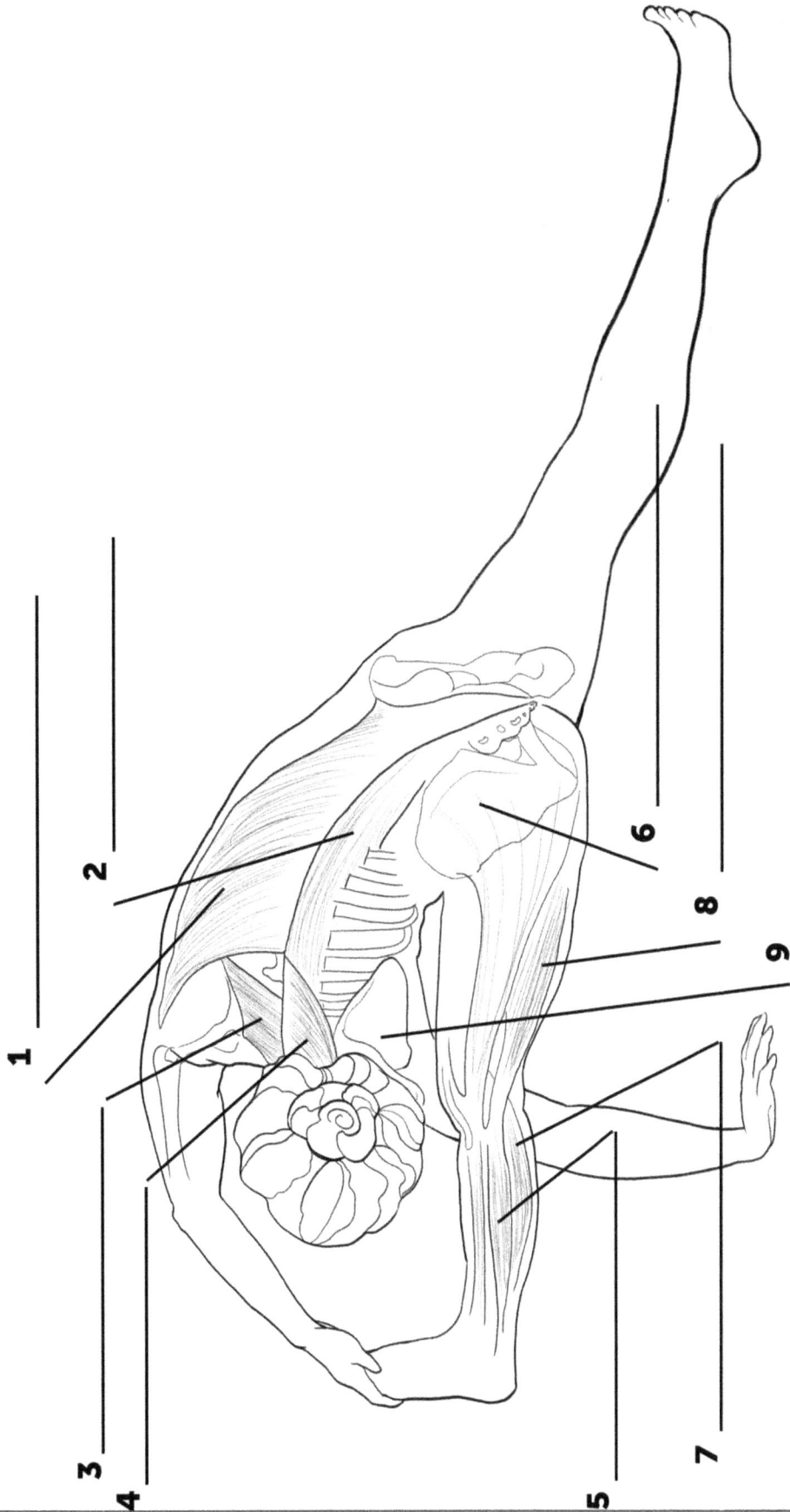

1

2

3

4

5

6

7

8

9

35. VISVAMITRASANA

1.   LATISSIMUS DORSI

2.   ERECTOR SPINAE

3.   RHOMBOIDS

4.   TRAPEZIUS

5.   SOLEUS

6.   PELVIS

7.   GASTROCNEMIUS

8.   HAMSTRINGS

9.   SCAPULA

36. BOUND SKANDASANA

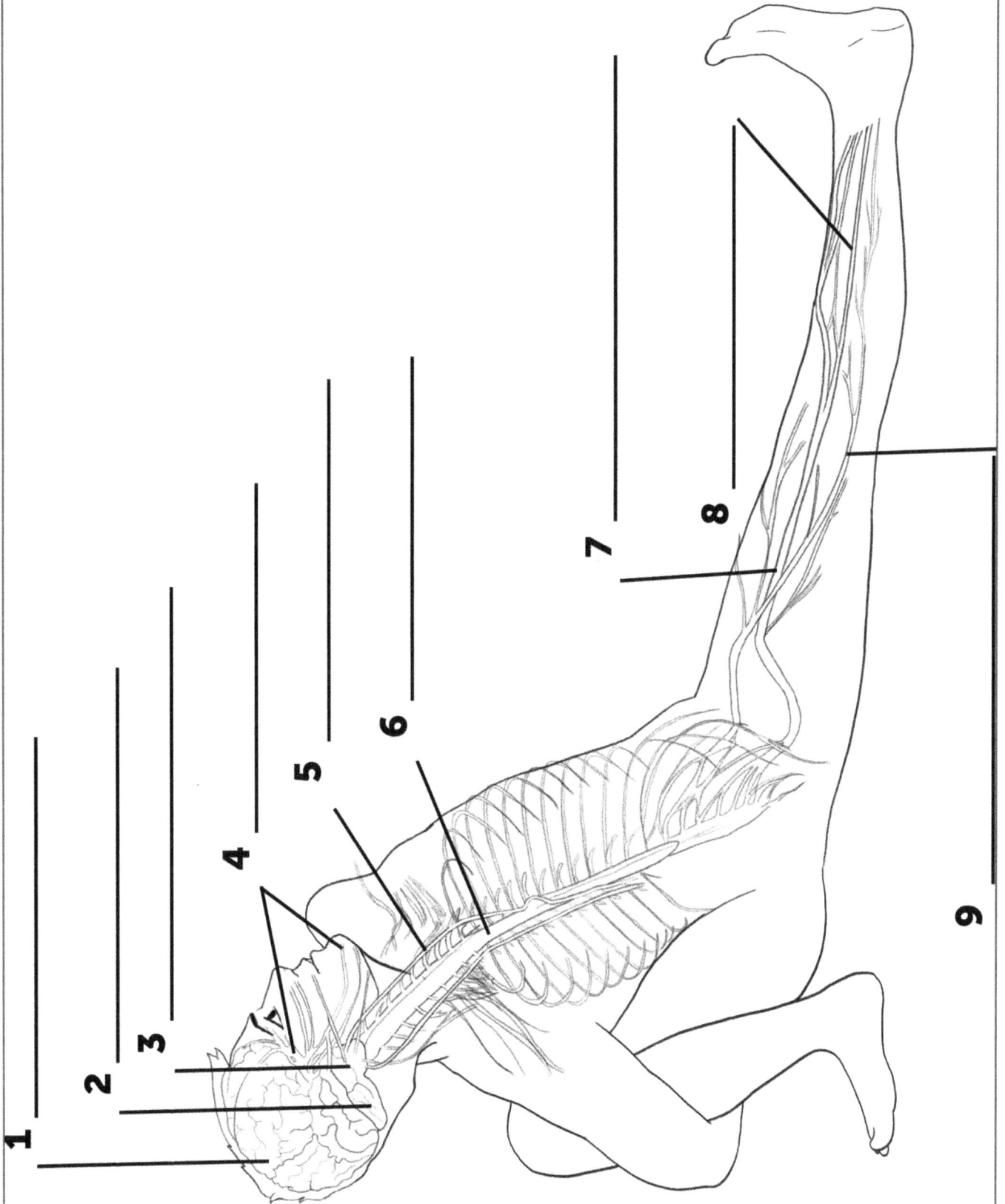

1

2

3

4

5

6

7

8

9

36. BOUND SKANDASANA

1. CEREBRUM
2. CEREBELLUM
3. BRAINSTEM
4. CRANIAL NERVES
5. VAGUS
6. SPINAL CORD
7. SCIATIC
8. TIBIAL
9. SAPHENOUS

37. DEVOTIONAL WARRIOR POSE

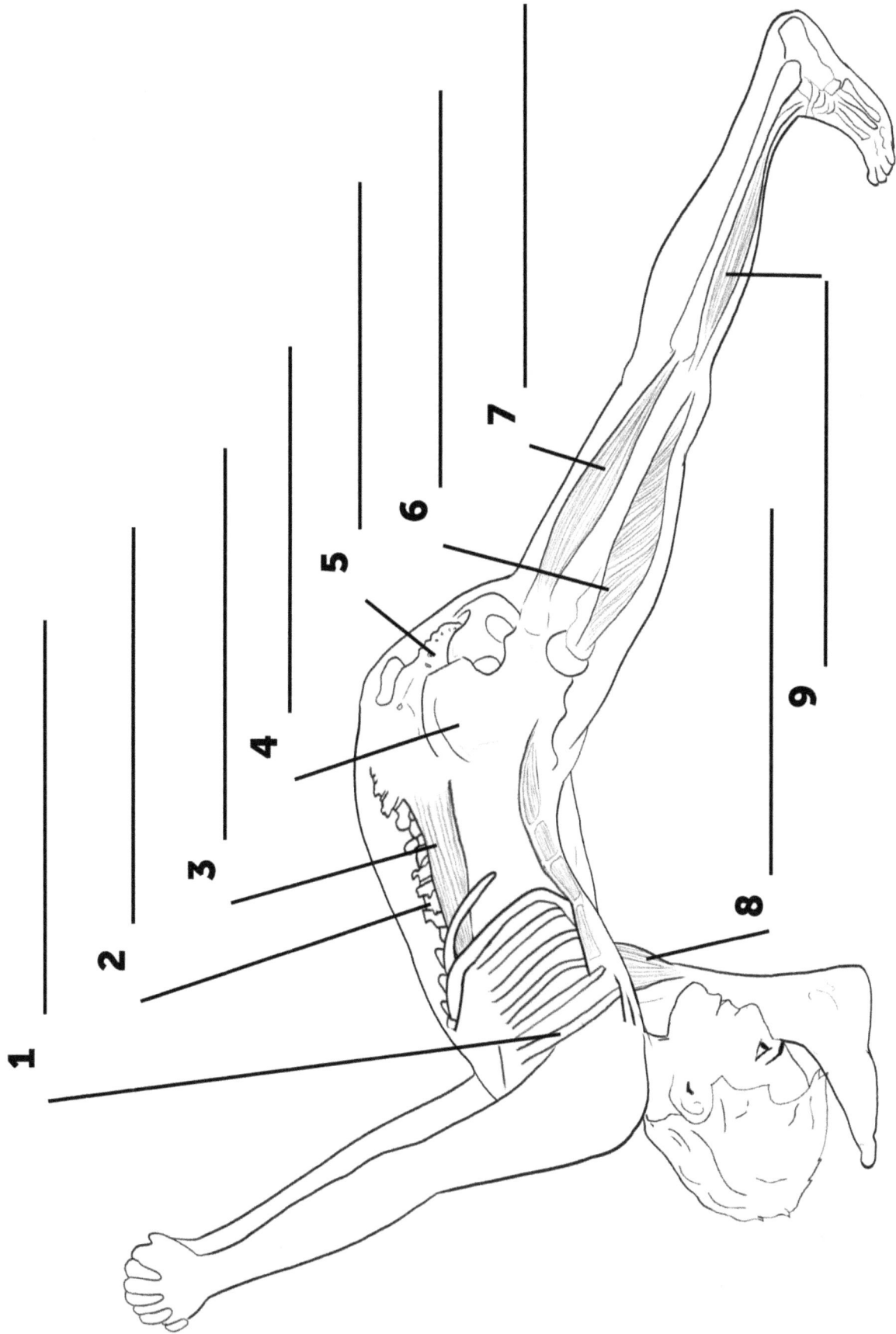

1

2

3

4

5

6

7

8

9

37. DEVOTIONAL WARRIOR POSE

1. RIBS
2. SPINE
3. ERECTOR SPINAE
4. PELVIS
5. SACRUM
6. QUADRICEPS
7. HAMSTRINGS
8. GASTROCNEMIUS
9. TIBIALIS ANTERIOR

38. BOUND LIZARD POSE

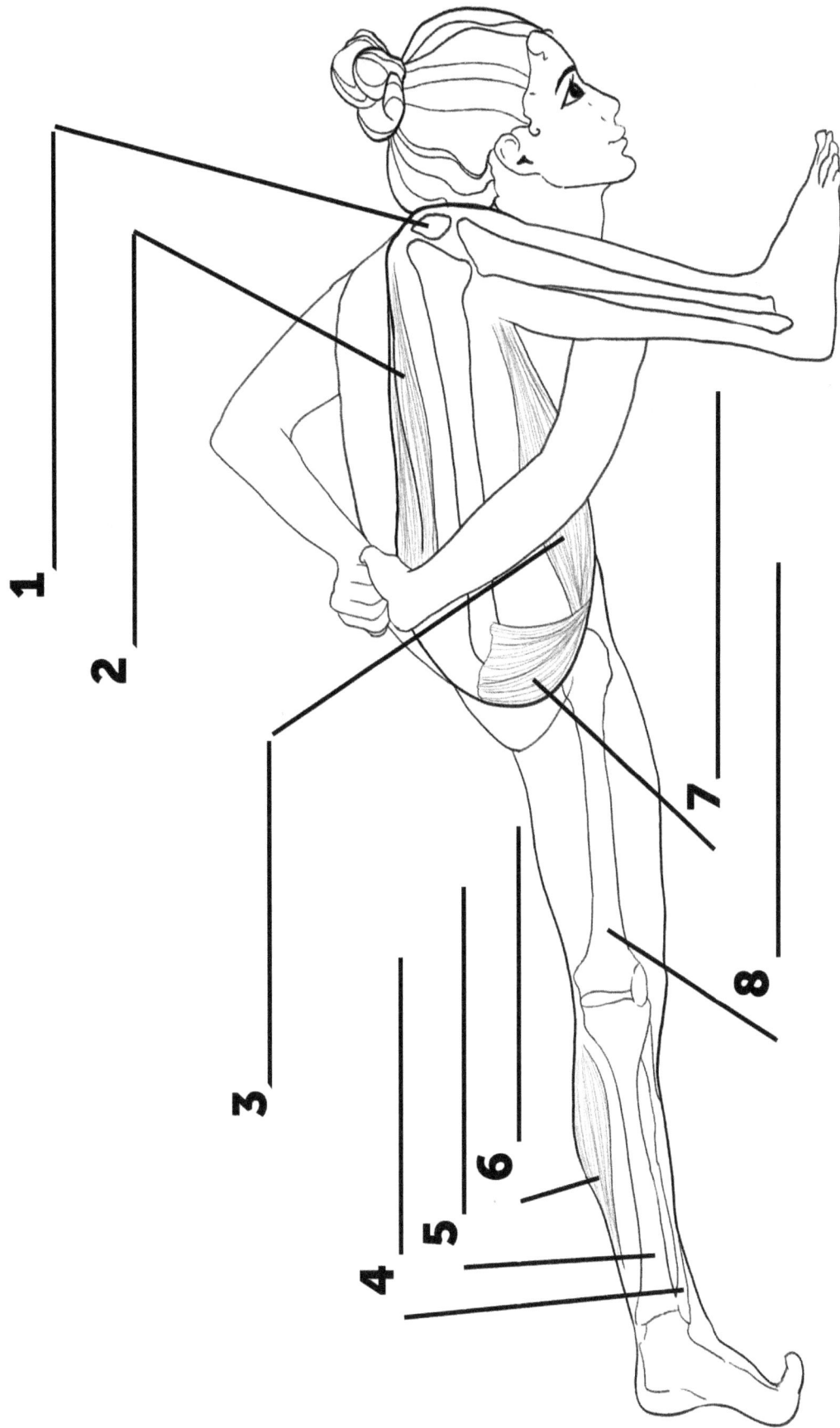

1

2

3

4

5

6

7

8

38. BOUND LIZARD POSE

1. PATELLA
2. QUADRICEPS
3. HAMSTRINGS
4. FIBULA
5. TIBIA
6. GASTROCNEMIUS
7. GLUTEUS MAXIMUS
8. FEMUR

39. STANDING SPLIT

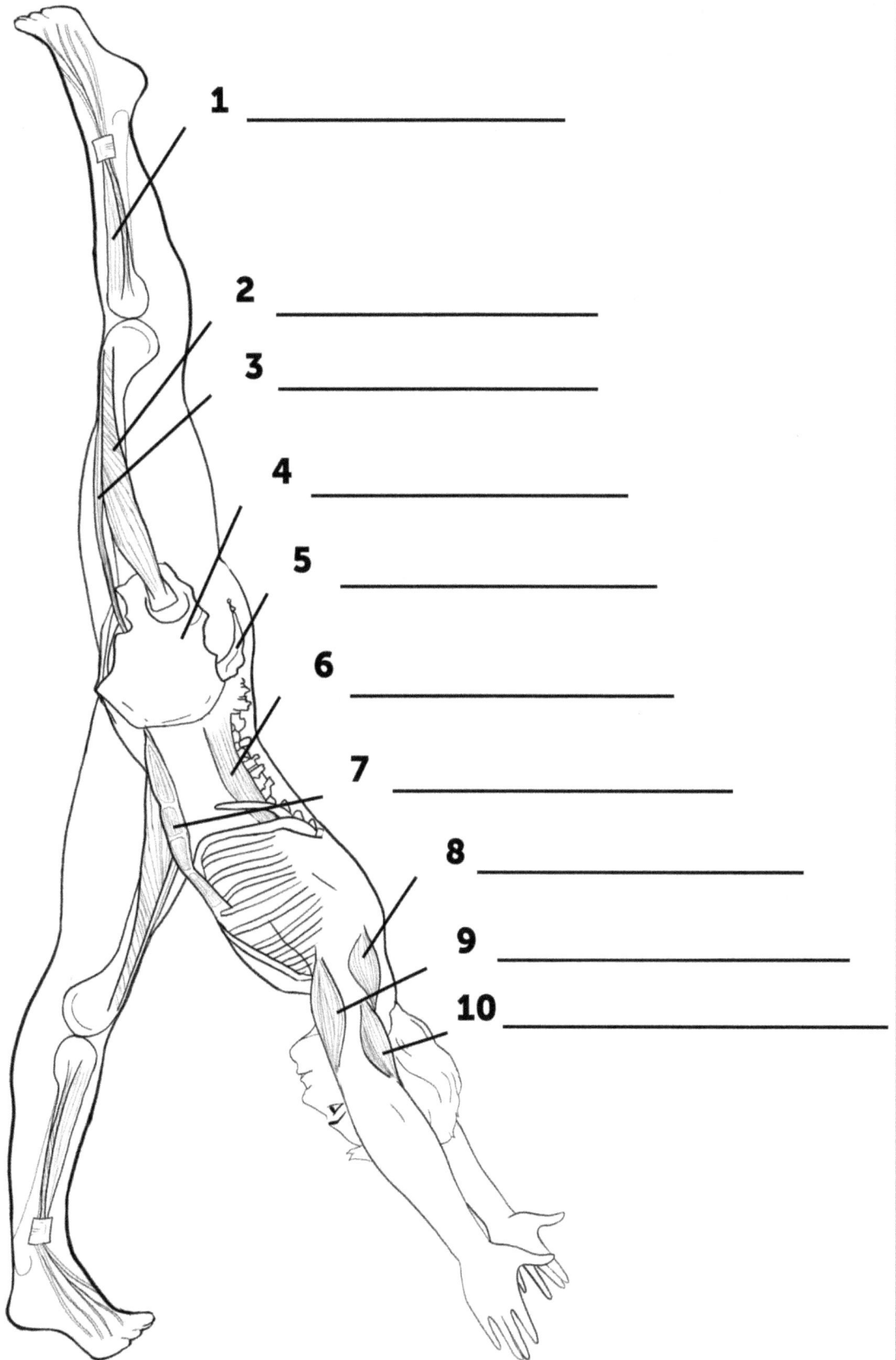

1 _____

2 _____

3 _____

4 _____

5 _____

6 _____

7 _____

8 _____

9 _____

10 _____

39. STANDING SPLIT

1. TIBIALIS ANTERIOR
2. RECTUS FEMORIS
3. SARTORIUS
4. PELVIS
5. SACRUM
6. ERECTOR SPINAE
7. RECTUS ABDOMINIS
8. DELTOID
9. BICEPS BRACHII
10. TRICEPS BRACHII

40. BOUND WARRIOR III

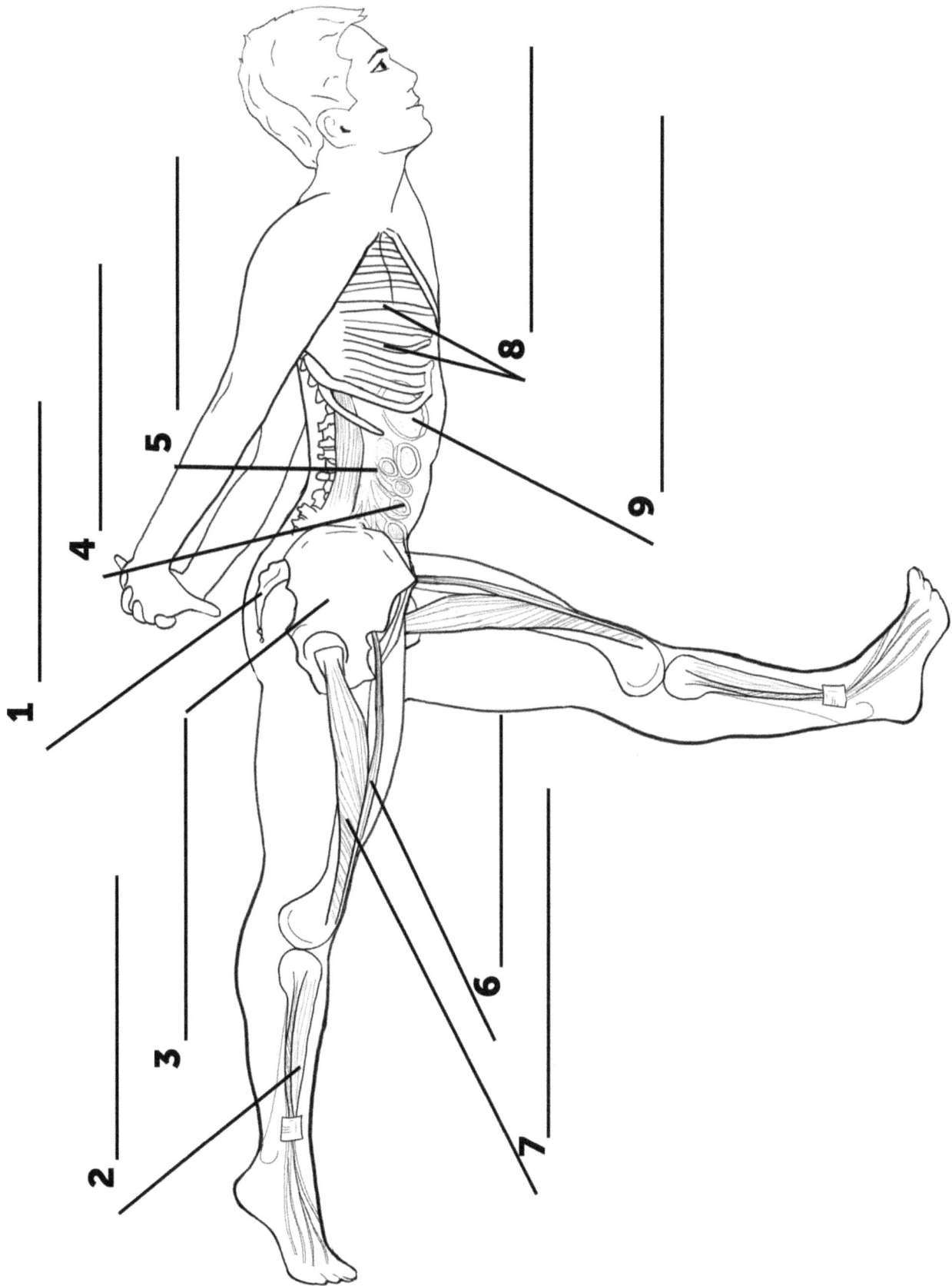

40. BOUND WARRIOR III

1. SACRUM
2. TIBIALIS ANTERIOR
3. PELVIS
4. COILS OF SMALL INTESTINE
5. MESENTERY OF SMALL INTESTINE
6. SARTORIUS
7. RECTUS FEMORIS
8. RIBS
9. STOMACH

41. BOUND FORWARD FOLD

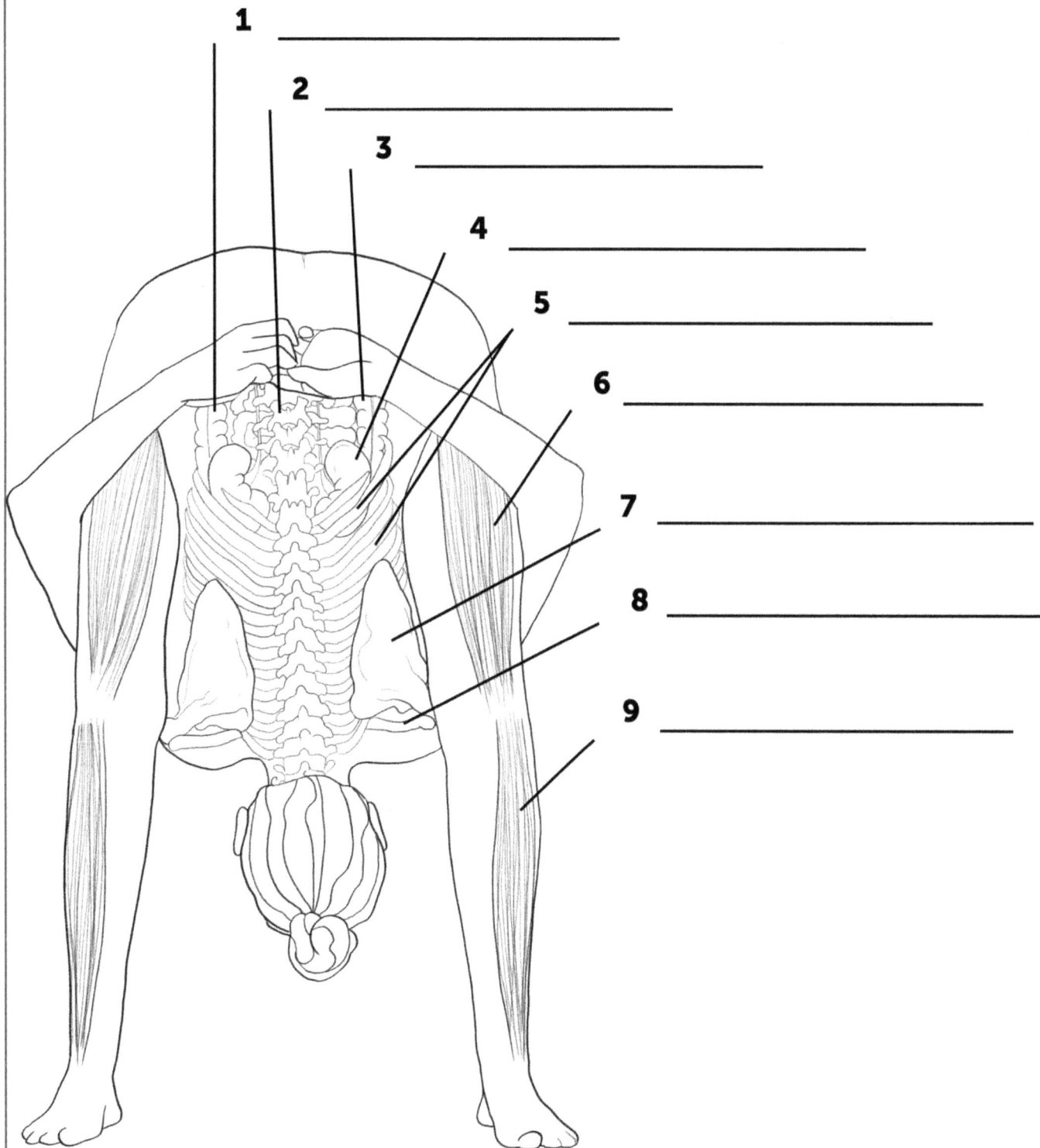

1 _____

2 _____

3 _____

4 _____

5 _____

6 _____

7 _____

8 _____

9 _____

41. BOUND FORWARD FOLD

1. ASCENDING COLON

2. SPINE

3. DESCENDING COLON

4. KIDNEY

5. RIBS

6. QUADRICEPS

7. SCAPULA

8. COLLARBONE

9. TIBIALIS ANTERIOR

42. RAG DOLL POSE

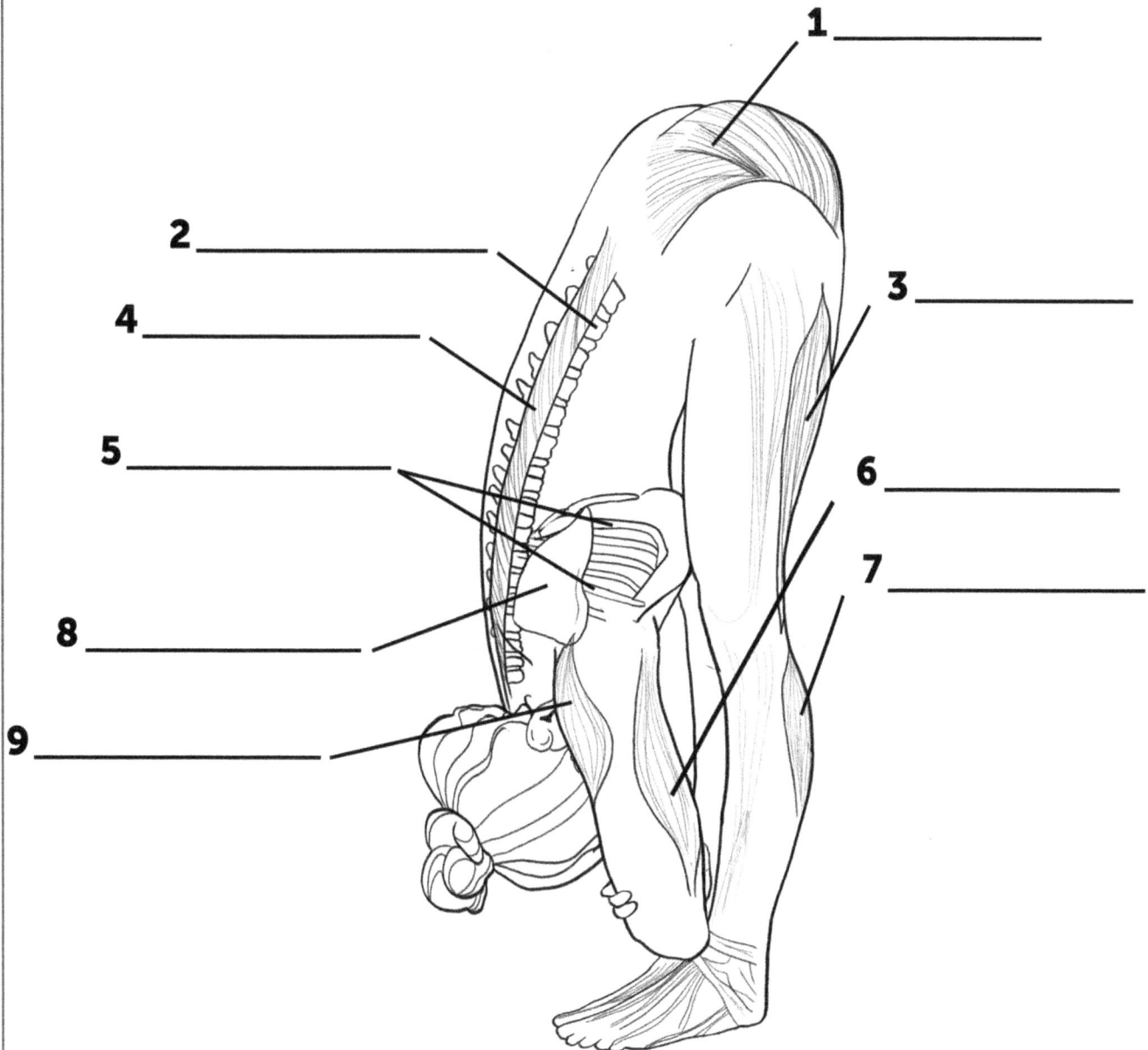

1_____

2_____

3_____

4_____

5_____

6_____

7_____

8_____

9_____

42. RAG DOLL POSE

1. PIRIFORMIS
2. SPINE
3. HAMSTRINGS
4. SPINAL MUSCLES
5. RIBS
6. TRICEPS BRACHII
7. GASTROCNEMIUS
8. SCAPULA
9. DELTOID

43. RESTED HALF PIGEON POSE

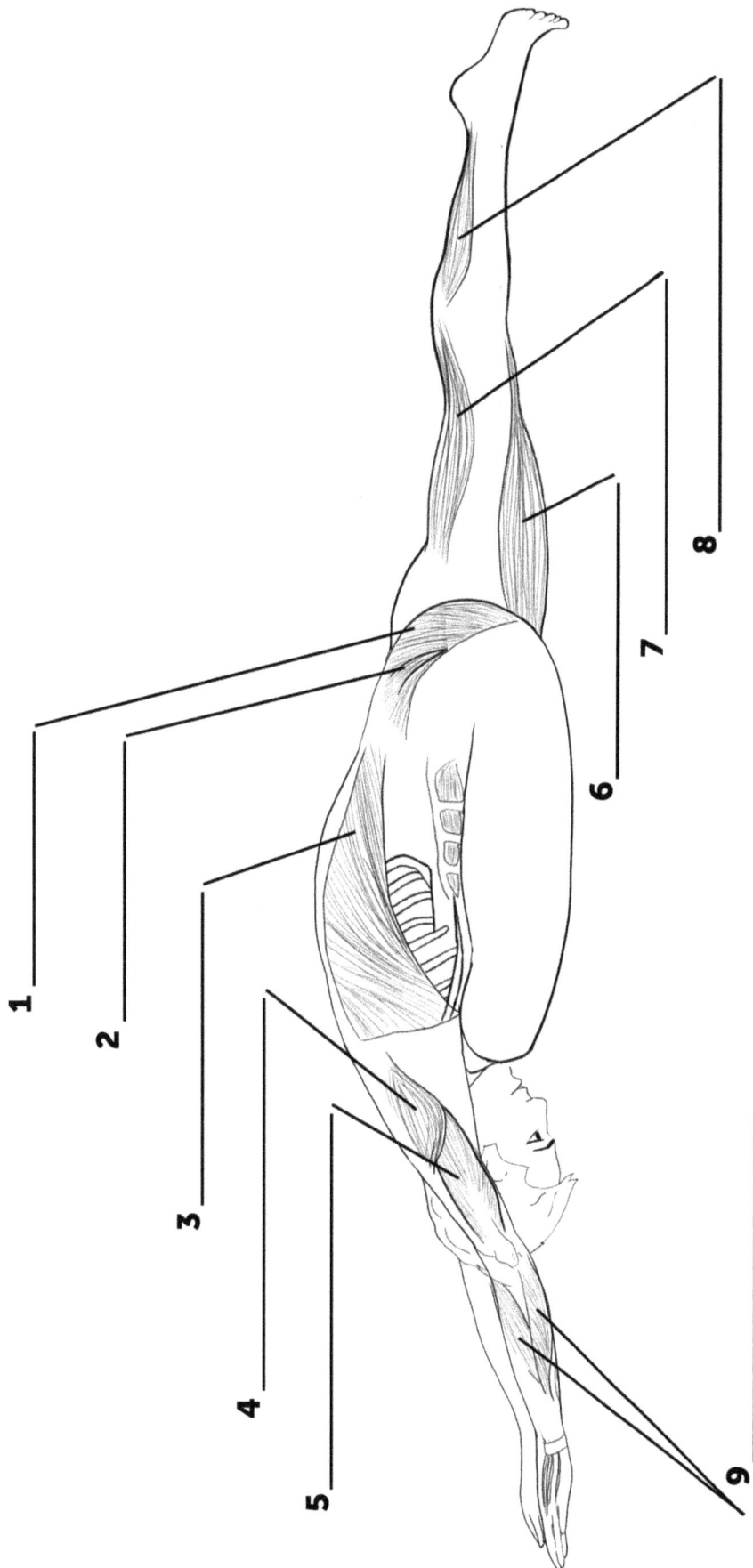

43. RESTED HALF PIGEON POSE

1. GLUTEUS MAXIMUS

2. PIRIFORMIS

3. LATISSIMUS DORSI

4. DELTOID

5. TRICEPS BRACHII

6. QUADRICEPS

7. HAMSTRINGS

8. GASTROCNEMIUS

9. PRONATORS

44. ONE LEGGED REVERSE TABLE

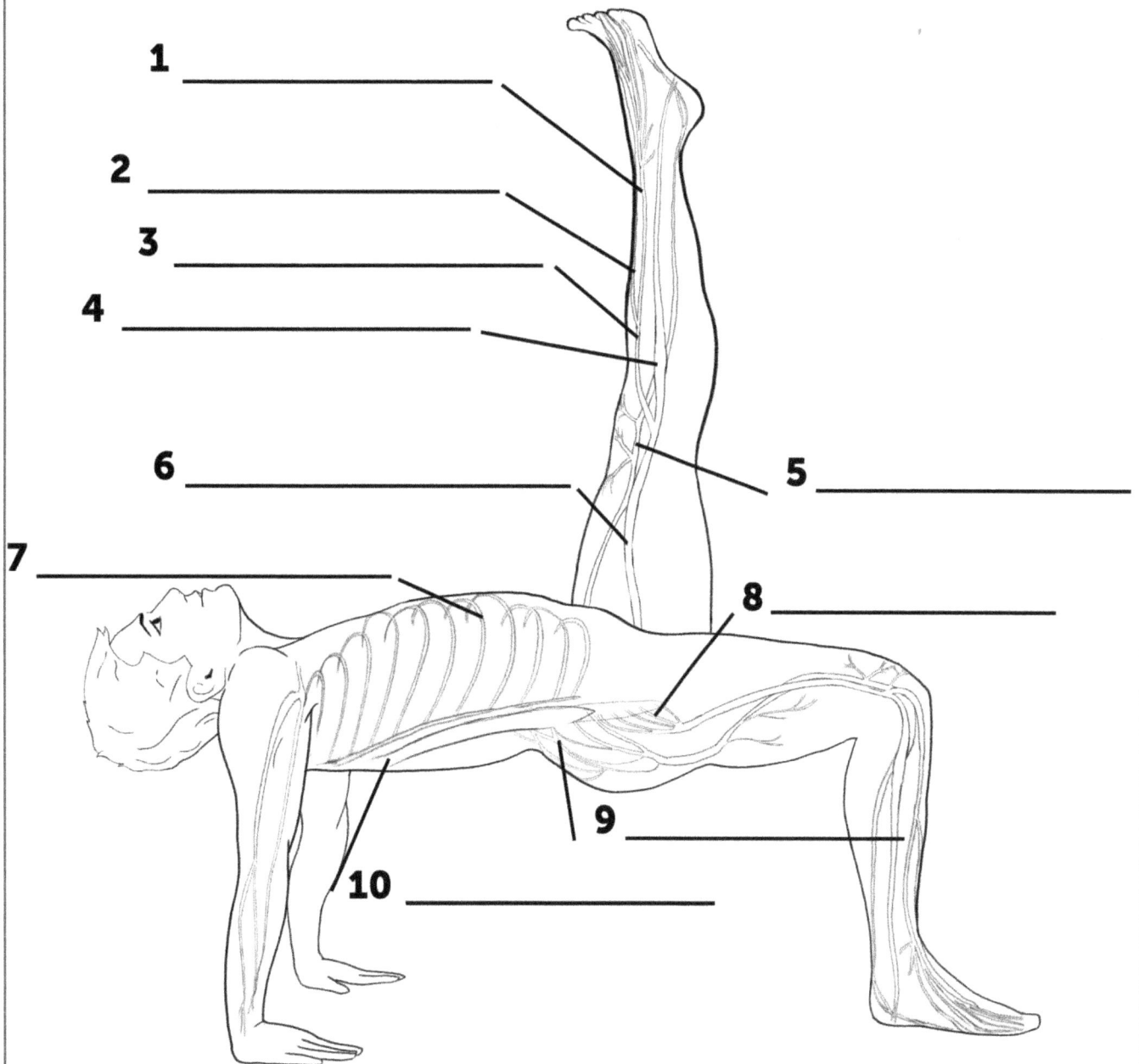

1 _____

2 _____

3 _____

4 _____

5 _____

6 _____

7 _____

8 _____

9 _____

10 _____

44. ONE LEGGED REVERSE TABLE

1. DEEP PERONEAL

2. SUPERFICIAL PERONEAL

3. COMMON PERONEAL

4. TIBIAL

5. SAPHENOUS

6. SCIATIC

7. INTERCOSTALS

8. SACRAL PLEXUS

9. LUMBAR PLEXUS

10. SPINAL CORD

45. ONE LEGGED CROW II

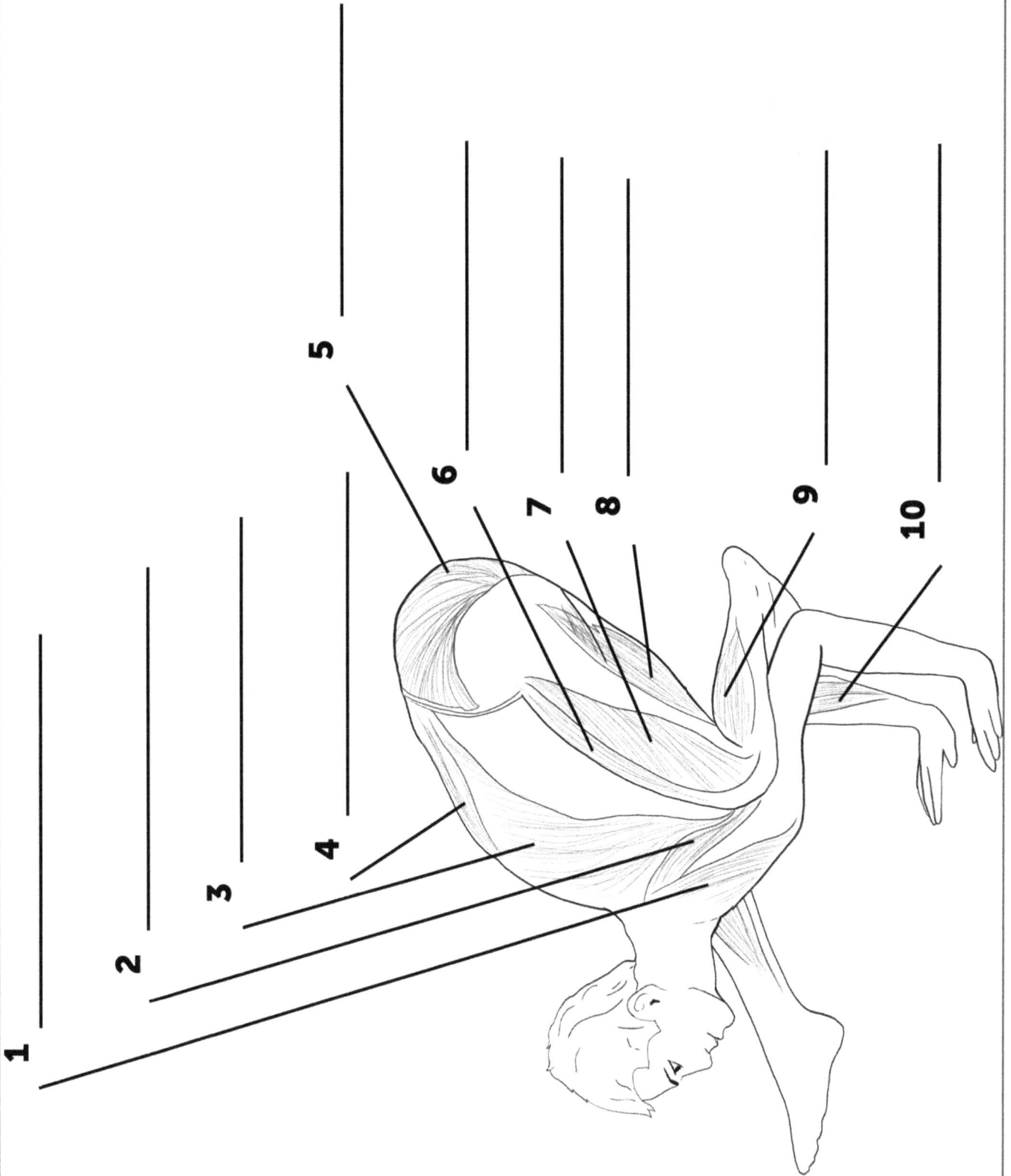

1

2

3

4

5

6

7

8

9

10

45. ONE LEGGED CROW II

1. DELTOID
2. TRICEPS BRACHII
3. LATISSIMUS DORSI
4. ERECTOR SPINAE
5. GLUTEUS MAXIMUS
6. RECTUS FEMORIS
7. VASTUS LATERALIS
8. HAMSTRINGS
9. GASTROCNEMIUS
10. PRONATORS

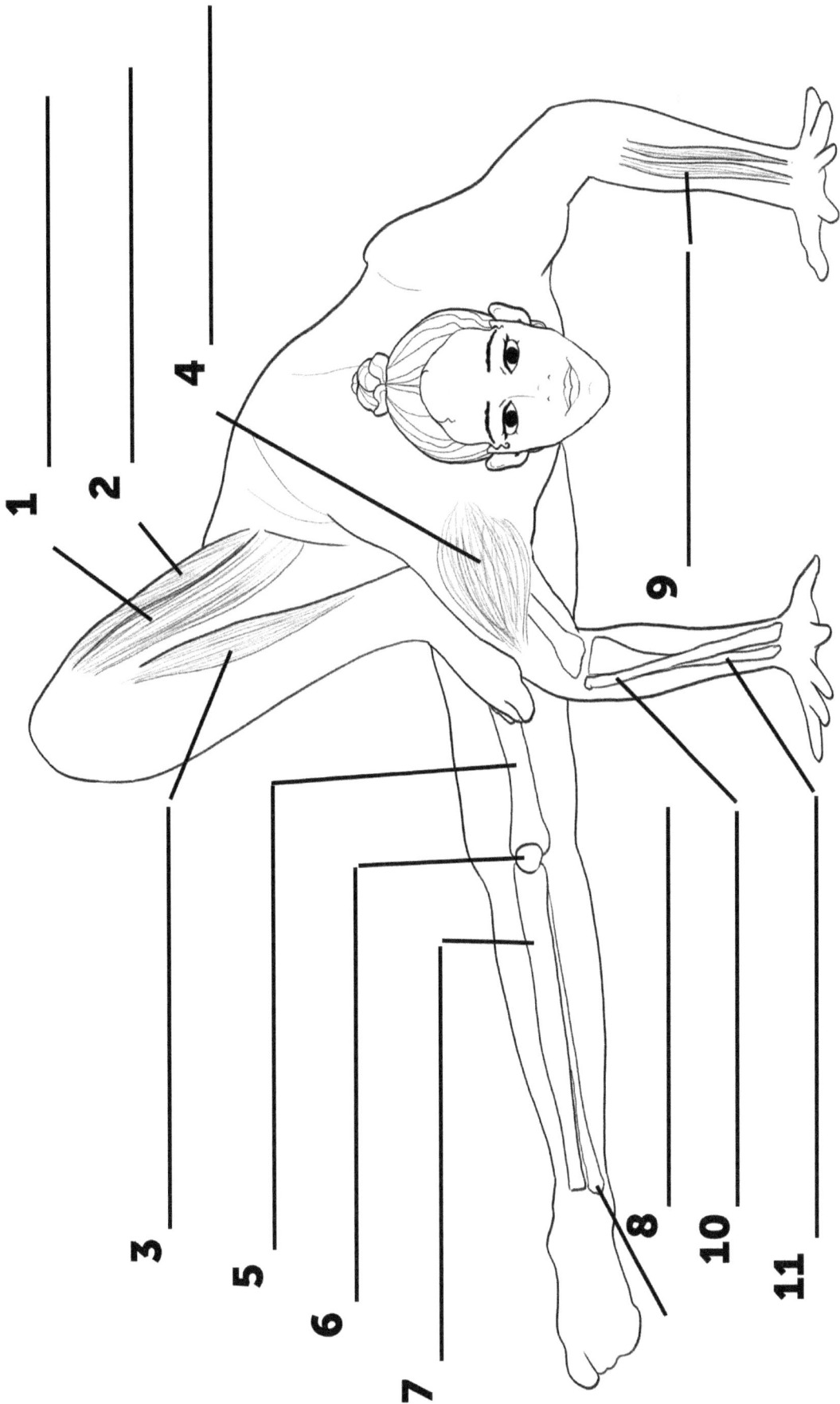

46. DRAGONFLY

1

2

3

4

5

6

7

8

9

10

11

46. DRAGONFLY

1. VASTUS LATERALIS
2. RECTUS FEMORIS
3. GASTROCNEMIUS
4. DELTOID
5. FEMUR
6. PATELLA
7. TIBIA
8. FIBULA
9. PRONATORS
10. RADIUS
11. ULNA

47. ONE HANDED TREE POSE

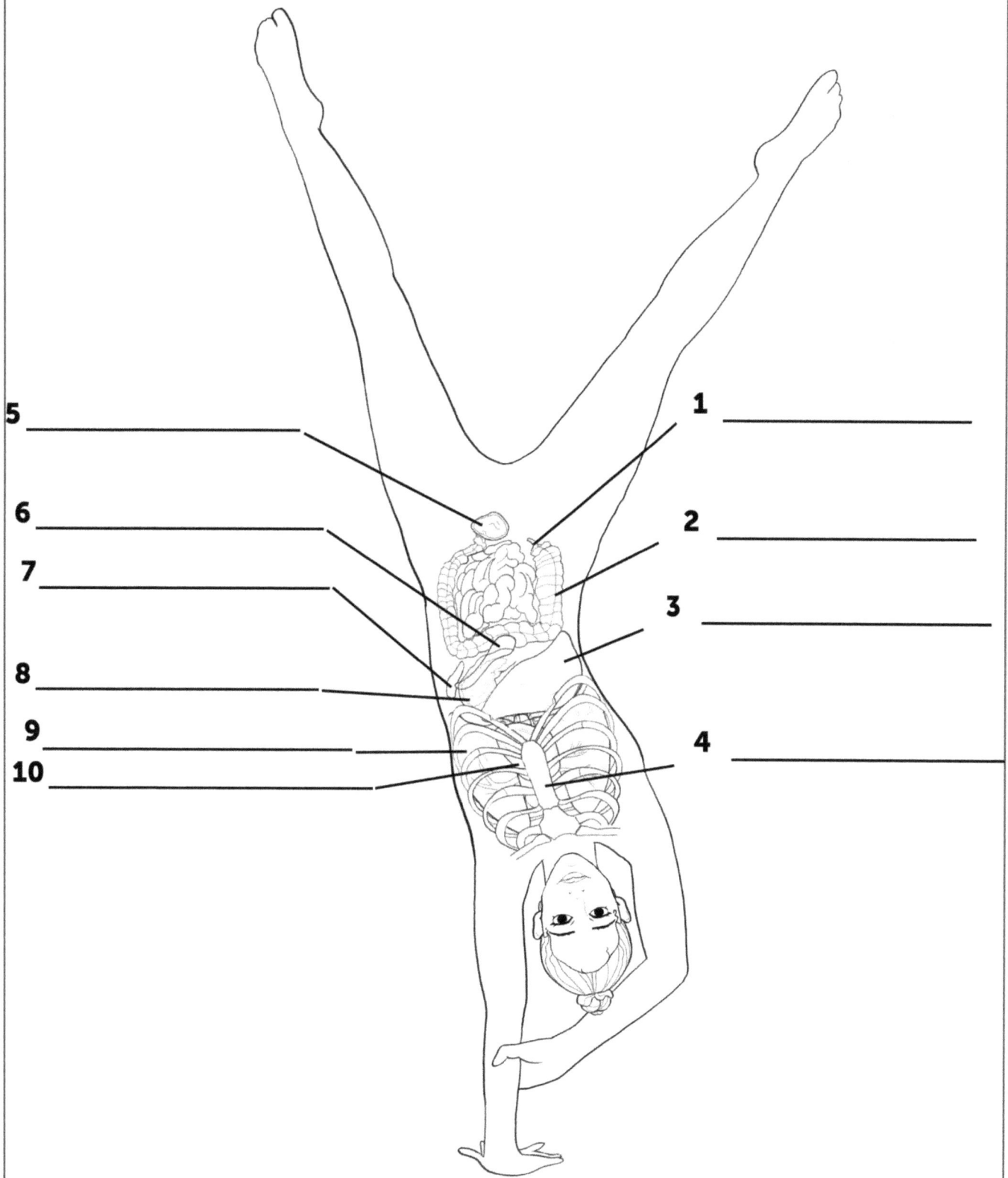

5 _____

6 _____

7 _____

8 _____

9 _____

10 _____

1 _____

2 _____

3 _____

4 _____

47. ONE HANDED TREE POSE

1. APPENDIX
2. ASCENDING COLON
3. LIVER
4. STERNUM
5. URINARY BLADDER
6. PANCREAS
7. SPLEEN
8. STOMACH
9. LUNGS
10. HEART

48. KING COBRA POSE

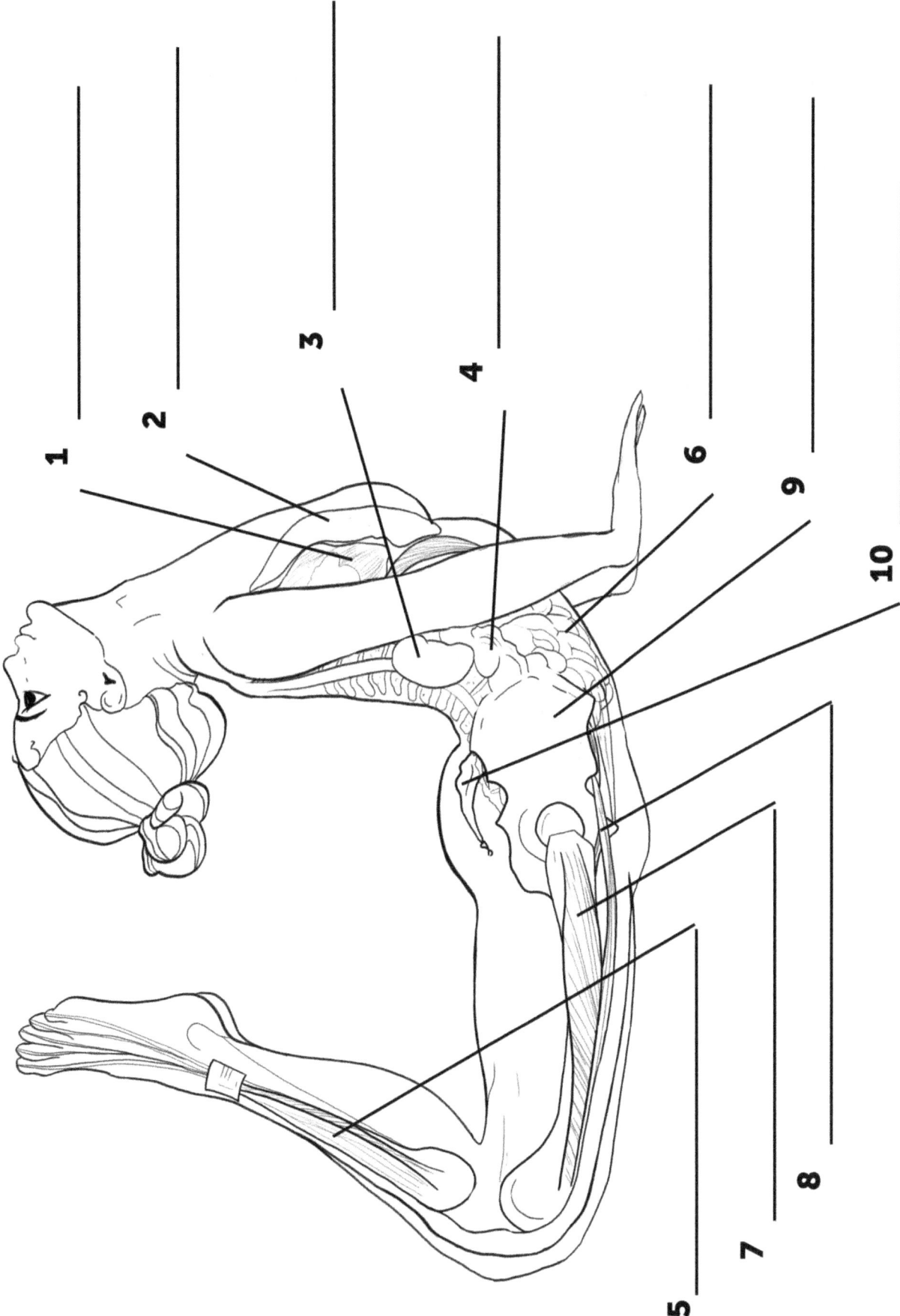

1

2

3

4

5

6

7

8

9

10

48. KING COBRA POSE

1. HEART
2. LUNGS
3. KIDNEY
4. ASCENDING COLON
5. TIBIALIS ANTERIOR
6. COILS OF SMALL INTESTINE
7. RECTUS FEMORIS
8. SARTORIUS
9. PELVIS
10. SACRUM

49. AWKWARD POSE

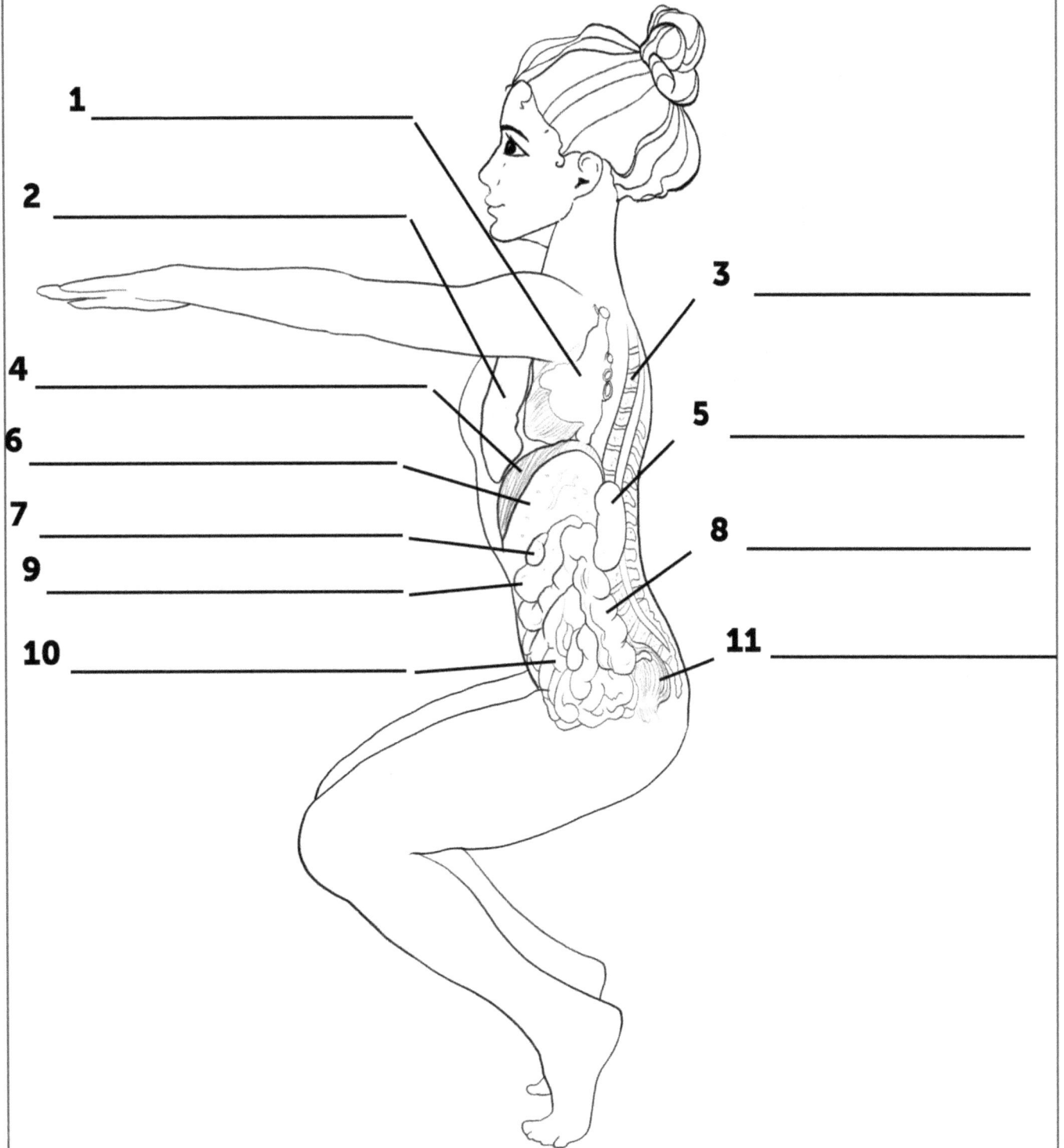

1 _____

2 _____

3 _____

4 _____

5 _____

6 _____

7 _____

8 _____

9 _____

10 _____

11 _____

49. AWKWARD POSE

1. HEART
2. LUNGS
3. SPINE
4. DIAPHRAGM
5. KIDNEY
6. LIVER
7. GALLBLADDER
8. DESCENDING COLON
9. STOMACH
10. COILS OF SMALL INTESTINE
11. RECTUM

50. STANDING HEAD TO KNEE

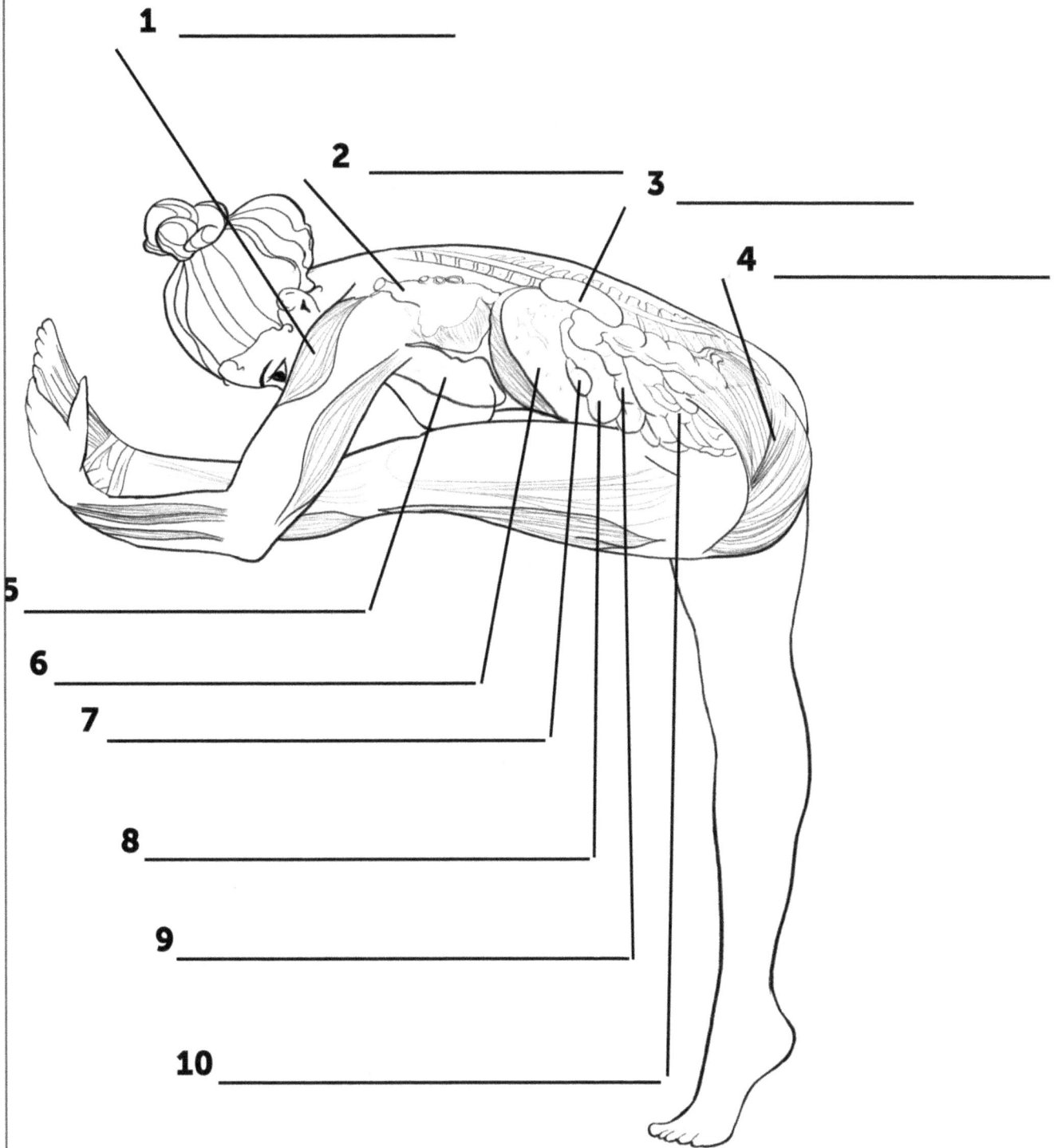

1 _____

2 _____

3 _____

4 _____

5 _____

6 _____

7 _____

8 _____

9 _____

10 _____

50. STANDING HEAD TO KNEE

1. DELTOID
2. HEART
3. KIDNEY
4. PIRIFORMIS
5. LUNGS
6. LIVER
7. GALLBLADDER
8. STOMACH
9. TRANSVERSE COLON
10. COILS OF SMALL INTESTINE

51. UNSUPPORTED SHOULDER STAND

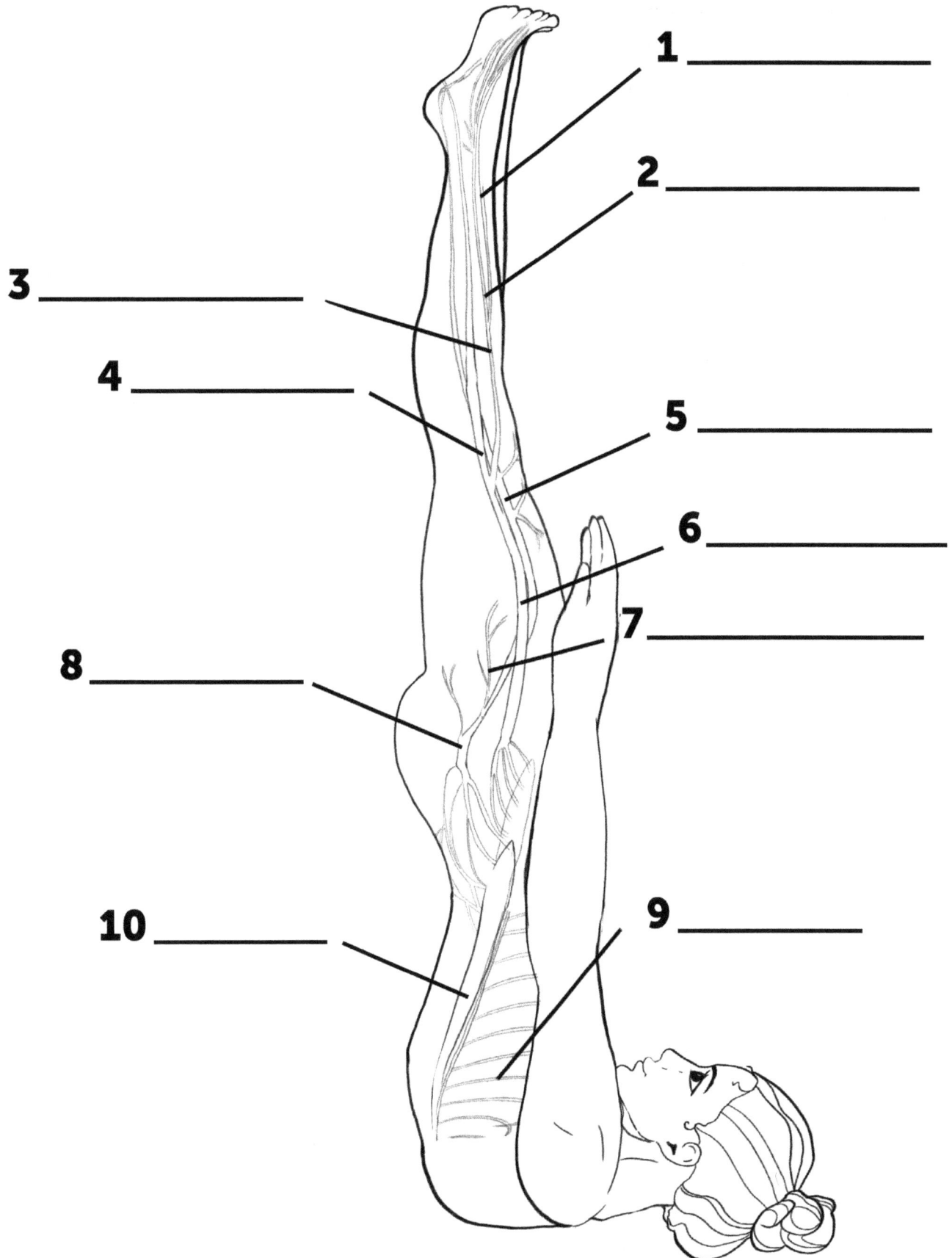

1 _____

2 _____

3 _____

4 _____

5 _____

6 _____

7 _____

8 _____

9 _____

10 _____

51. UNSUPPORTED SHOULDER STAND

1.   SUPERFICIAL PERONEAL

2.   DEEP PERONEAL

3.   COMMON PERONEAL

4.   TIBIAL

5.   SAPHENOUS

6.   SCIATIC

7.   MUSCULAR BRANCHES OF FEMORAL

8.   FEMORAL

9.   INTERCOSTALS

10.   SPINAL CORD

52. SKANDASANA

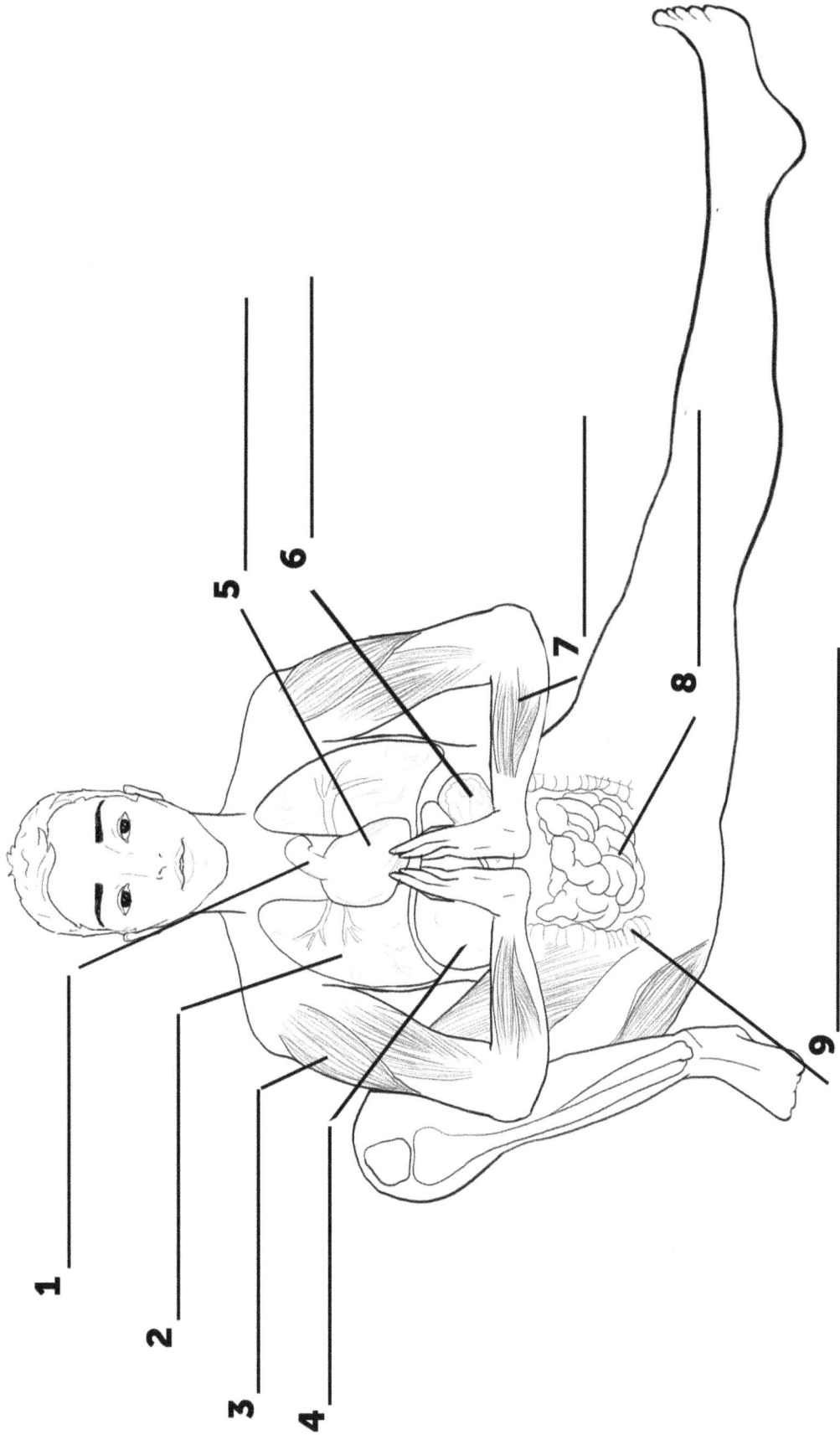

1

2

3

4

5

6

7

8

9

52. SKANDASANA

1. AORTA
2. LUNGS
3. DELTOID
4. LIVER
5. HEART
6. STOMACH
7. PRONATORS
8. COILS OF SMALL INTESTINE
9. ASCENDING COLON

53. SIDE-RECLINING LEG LIFT

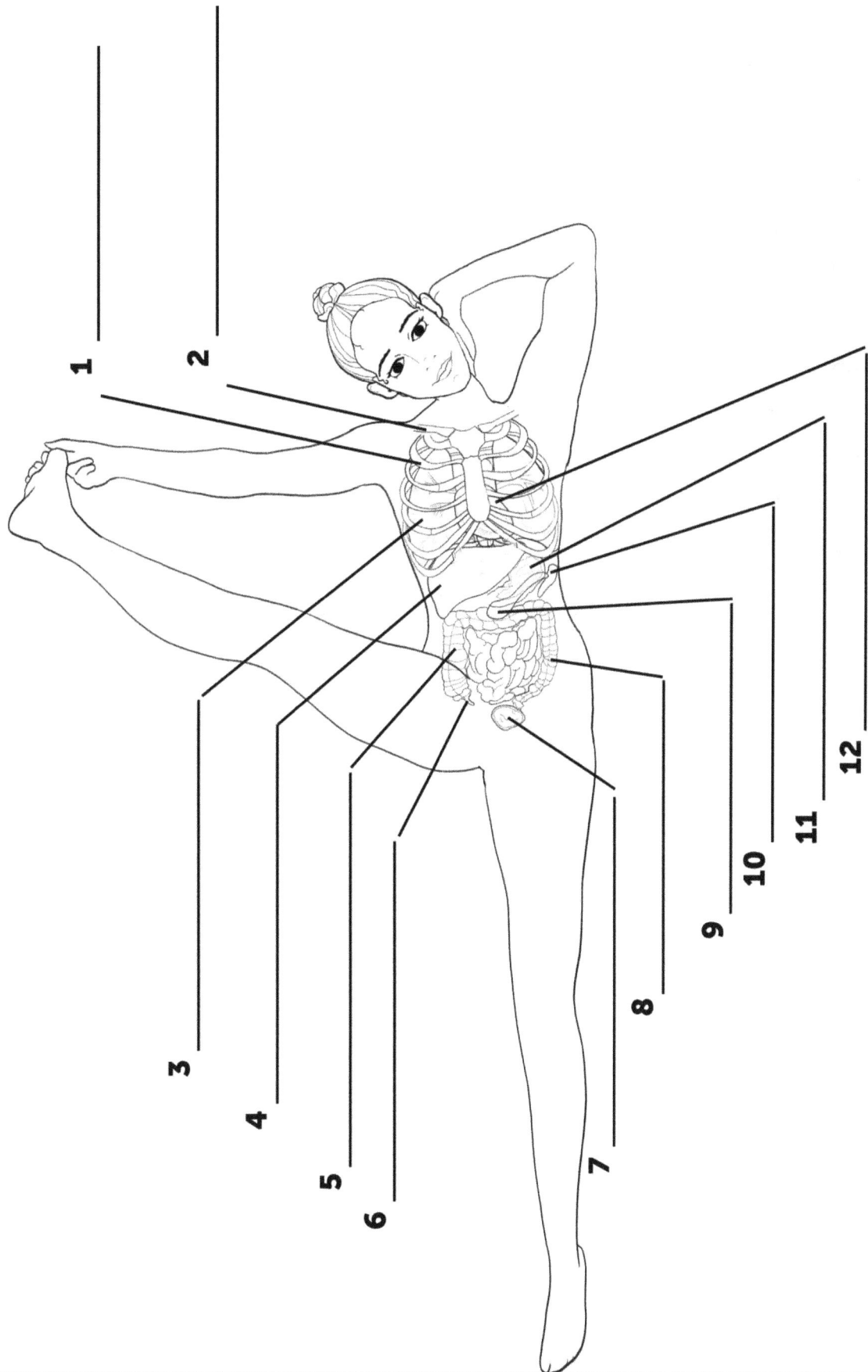

1

2

3

4

5

6

7

8

9

10

11

12

53. SIDE-RECLINING LEG LIFT

1. RIBS
2. COLLARBONE
3. LUNGS
4. LIVER
5. ASCENDING COLON
6. APPENDIX
7. URINARY BLADDER
8. DESCENDING COLON
9. PANCREAS
10. SPLEEN
11. STOMACH
12. HEART

www.ingramcontent.com/pod-product-compliance
Lightning Source LLC
Chambersburg PA
CBHW051342200326
41521CB00015B/2587